FEEDBACK MECHANISMS
IN ANIMAL BEHAVIOUR

FEEDBACK MECHANISMS IN ANIMAL BEHAVIOUR

D. J. McFARLAND

Balliol College
Oxford, England

1971

Academic Press · London and New York

Academic Press Inc. (London) Ltd.
24-28 Oval Road
London NW1 7DX

U.S. Edition published by
ACADEMIC PRESS INC.
111 Fifth Avenue
New York, New York 10003

Library of Congress Catalog Card Number: 78-170755
ISBN: 0.12.483850.2

Printed in Great Britain by
The Whitefriars Press Ltd., London and Tonbridge

Preface

Although it is a truism to say that behaviour is a function of time this aspect of behaviour is by no means the only, or even the most common, focal point for behaviour study. In studying evolutionary or genetic aspects of behaviour, for instance, the way in which behaviour changes from one generation to the next is of prime importance. Similarly, in learning studies, behaviour is generally considered as a function of trials, or some other measure of experience. In studying behaviour as a function of time, one is primarily concerned with the performance of particular behaviour patterns, their orientation, goal directiveness, and motivation. It is with these aspects of behaviour that this book is primarily concerned.

The study of behaviour as a function of time is partly concerned with the way behaviour is controlled by internal and environmental stimuli. The emphasis, in this book, is on the control aspect of this problem. An attempt is made to introduce students of animal behaviour to some of the more elementary aspects of the theory of control, as developed by physical scientists. Control theory is a discipline which cuts across traditional boundaries between scientific fields of study and uses the same language for systems of different hardware, whether physical or biological.

Intended to give an understanding of control theory that is essentially intuitive, this book attempts to keep the mathematics to a minimum and to bypass certain of the more advanced aspects, such as complex numbers, which are normally considered essential for certain branches of control theory. The mathematically minded will therefore notice a certain lack of rigour: this is intentional. Once the biologist has an understanding of what control theory is about, and how it is relevant to his problems, he can readily move on to one of the many excellent texts that are already available.

Though not primarily a text on control theory, the book is

organized on the basis of aspects of control, arranged in order of increasing complexity. A consequence of this organization is that discussion of behaviour is used to illustrate the basic control theme, rather than to illustrate points about behaviour itself. Thus behavioural examples have been used where they are suitable for the illustration of particular aspects of control theory. Much excellent work relating particularly to the control of invertebrate behaviour is omitted, simply because it is too complex or too specialized to be used. In addition, a few selected examples are used repeatedly, and elaborated a little each time. Nowhere is a high degree of biological knowledge assumed. It is hoped, therefore, that the book will prove palatable to biologists interested in control and to physical scientists interested in behaviour.

The application of control theory to behaviour does raise some philosophical issues relating to the nature of explanation in behaviour. These are not specifically dealt with here, though a number of suggestions are made concerning the manner in which control concepts might be used in the description and explanation of behaviour. Similarly, there is a certain amount of comment on the ways in which control concepts have been used, or misused, in the past. This, it is hoped, will provide a suitable basis for future discussion of problems of explanation.

One such problem concerns the question of whether behaviour should ultimately be explained in physiological terms. Control theory offers a type of explanation that can be applied to the study of systems in general, be they physical, biological or economic. Such explanation is independent of the hardware of the system concerned, and is a truly behavioural type of explanation. In relation to animal behaviour, this type of explanation is abstract in the sense of being independent of the particular physiological processes responsible for the behaviour. This does not mean that there is necessarily any intrinsic merit in the hardware–independent approach, as opposed to the study of the "real nervous system" (Weiskrantz, 1968), or vice versa. Both types of explanation are necessary and neither is sufficient. The reader of this book will discover that there are problems of control that can only be solved by physical verification. There are also problems of brain function that can be solved only through greater knowledge of behaviour. At the empirical level the two approaches can be complementary. For example, removal of a particular control function, by a particular brain treatment, can be used as evidence in a theory of control, even though the physiological effects of the treatment may be unknown. Conversely, differences in control of behaviour can be used as evidence in the investigation of brain function.

This book would not have been possible without the many friends and colleagues who have assisted with my, often painfully slow, education. In particular, I am grateful to Niko Tinbergen and David Vowles for encouraging my initial ventures into the realm of feedback theory: to Ernie Freeman and David Witt for my first expert advice on control theory. My thanks go to my colleagues Philip Budgell, Juan Delius and Ivor Lloyd for our many stimulating discussions and to all who have kindly allowed me to make use of their work. I am especially indebted to Phil Budgell, Mike Cullen, Oliver Jacobs and Jill McFarland for reading the manuscript and for their helpful comments. Many authors have kindly allowed me to reproduce figures from their work and these are acknowledged in the legends. I am also indebted to the following publishers for permission to reproduce material: Little, Brown and Company; McGraw-Hill Book Company; Plenum Press; W. B. Saunders Company; John Wiley and Sons.

October 1971   D. J. McFARLAND
                                        Balliol College and
                          Department of Experimental
                                  Psychology, Oxford

Contents

To N. T. and D. M. V.

CHAPTER 1

An Outline of Control Theory

Most biologists are familiar with the notion that control over the behaviour of variables can be exercised by means of feedback mechanisms. The idea of control was implicit in Claude Bernard's concept of the constancy of the internal environment, though the explicit postulate of the role of nervous and chemical feedback mechanisms in the maintenance of physiological steady-state was formulated later, and embodied in Walter Cannon's concept of homeostasis. Feedback processes are now recognized to occur in the regulation of muscular movement, the control of metabolism and a host of other biological functions.

In the physical sciences, the feedback concept can be traced back to the thermostatically controlled furnace invented by Drebbel (1573-1633) and the centrifugal governor invented by Huygens (1629-1695) for use in clocks. It was adapted for windmills and water-wheels before being used by James Watt for the steam engine. James Clerk Maxwell (1868) was the first to analyse such phenomena in mathematical terms and can be regarded as the father of modern control theory.

Control theory is a discipline which cuts across traditional boundaries between fields of scientific study and uses the same language for systems of different hardware, whether electrical, mechanical, or biological. The application of control theory to biological problems stems from the recognition that analogies can exist between the behaviour of physical and biological systems. The subject of analogies thus serves as a useful introduction to the nature of control theory.

1.1 ANALOGIES

Analogies between various laws of nature have proved important in the development of the physical sciences, particularly in the application of familiar concepts to fresh fields of study, as instanced by the work of Georg Simon Ohm (1787-1854), who looked upon the flow of electric current in a wire as analogous to the flow of heat along a conductor.

Two systems can be said to be analogous when their behaviour, defined by an equation, is identical. As an example, consider the mechanical system portrayed in Fig. 1.1a. A mass M is suspended in a

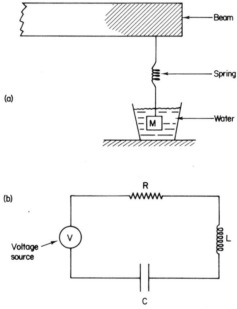

Fig. 1.1. Analogous (a) mechanical and (b) electrical systems.

bucket of water from a spring of stiffness K attached to a fixed beam. Applying Newton's law, force equals mass times acceleration, we can write

$$f_m = M\frac{du}{dt} \qquad (1.1)$$

The force needed to push the mass through the water is proportional to the velocity of the mass. Thus our second equation is

$$f_b = Bu \qquad (1.2)$$

where B is the constant of proportionality, called the friction. The third equation relates the rate of change of displacement of the spring (i.e. the spring velocity) to the applied force.

$$f_k = \frac{1}{K} \int u \, dt \tag{1.3}$$

where K is a constant determining the compliance of the spring. Suppose that no external force is applied after the mass is initially set in motion: what will be the behaviour of the system? According to the laws of mechanics the algebraic sum of the forces applied to a body is zero. In the present case

$$f_m + f_b + f_k = 0 \tag{1.4}$$

Substituting eqns. (1.1), (1.2) and (1.3), the behaviour of the system can be described by a single equation

$$M \frac{du}{dt} + Bu + \frac{1}{K} \int u \, dt = 0 \tag{1.5}$$

Now, consider the electrical circuit portrayed in Fig. 1.1b. A resistance, inductance and a capacitance are wired in series with a voltage source. The voltage across the inductance is given by the equation

$$v_1 = L \frac{di}{dt} \tag{1.6}$$

the voltage across the resistance is given by Ohm's law

$$v_r = Ri \tag{1.7}$$

and the voltage across the capacitance is given by the equation,

$$v_c = \frac{1}{C} \int i \, dt \tag{1.8}$$

According to Kirchoff's voltage law the algebraic sum of all the voltages around a closed-loop is zero. Thus if the source is suddenly dropped to zero, the behaviour of the system can be described by the equation

$$v_1 + v_r + v_c = 0 \tag{1.9}$$

Substitution of eqns. (1.6), (1.7) and (1.8) yields the following equation

$$L \frac{di}{dt} + Ri + \frac{1}{C} \int i \, dt = 0 \tag{1.10}$$

By inspection, eqns. (1.5) and (1.10) can be seen to be similar. In fact, if we substitute u for i, M for L, B for R, and K for C, they become identical. We can justifiably say that the two systems portrayed in Fig. 1.1 are analogous and we can draw up a table of analogous variables and parameters, such as Table 1.1. Analogies such as this are well known in the physical sciences and have been extended to acoustic, hydraulic and thermal systems.

The existence of analogies makes it possible to define "generalized" system variables and parameters. Thus, as illustrated in Table 1.1, the analogy between force and voltage leads to the

Table 1.1. Analogous variables and parameters

	Mechanical	Electrical	General
Rate Variables	force f velocity u	voltage v current i	effort e flow f
State Variables	displacement x momentum p	charge q flux λ	displacement h momentum p
Parameters	mass M compliance K friction B	inductance L capacitance C resistance R	inductance L capacitance C resistance R
Power	$HP = fu$	$P = vi$	$P = ef$
Potential energy	$E_p = \frac{1}{2}Kf^2$	$E_p = \frac{1}{2}Cv^2$	$E_p = \frac{1}{2}Ce^2$
Kinetic energy	$E_k = \frac{1}{2}Mu^2$	$E_k = \frac{1}{2}Li^2$	$E_k = \frac{1}{2}Lf^2$

concept of a generalized "effort" variable e. Similarly, the generalized "flow" variable f comes from the velocity-current analogy. In terms of the generalized variables and parameters, listed in Table 1.1, eqns. (1.5) and (1.10) can be rewritten as follows

$$L\frac{df}{dt} + Rf + \frac{1}{C}\int f dt = 0 \qquad (1.11)$$

This equation summarizes the behaviour of this type of system in the classical manner. However, in terms of control theory the nature of the system can also be described in a topological fashion.

1.2 BLOCK DIAGRAMS

The functional relationship between the variables of a system can conveniently be described in terms of a block diagram, in which the parameters of the system are represented by boxes, and the independent and dependent variables by arrows pointing respectively into and out of each box. For example, the three basic elements of the system described by eqn. (1.11) are portrayed in block diagram form in Fig. 1.2a. A block diagram of the whole system can be constructed by a suitable combination of the elements, as shown in Fig. 1.2b.

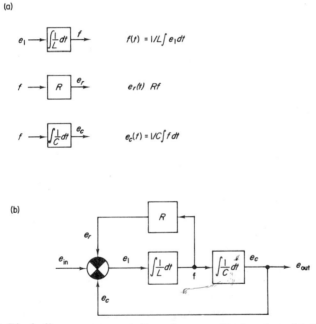

Fig. 1.2. Block diagram representation of a generalized system: (a) the system components, (b) the synthesized system.

An essential feature of the block diagram description of a system is the distinction made between variables and parameters. It is important that this distinction should be appreciated. In general, variables represent those measurable properties of the system which change value during the period of observation or experiment. In the system portrayed in Fig. 1.1b the current flowing in the wire is a variable, as it can change value from one moment to the next.

Parameters can be taken as those properties of the system which do not change during the specified period. Thus, the value of the electrical resistance in Fig. 1.1b is a parameter and appears as a constant in the system equation. Unfortunately this distinction is not a rigid one, as it is possible for parameters to change value during the specified period. This could happen if the resistance in Fig. 1.1b were replaced by a variable resistance or potentiometer. In such a case the value of the resistance would be regarded, not as a variable, but as a changing parameter. The coefficients of the differential equation describing the system would not be constant and the system would be said to be time-varying.

Fig. 1.3.

The role of parameters in a system may be easier to appreciate if the relationship between variables is viewed as a cause-and-effect relationship, as shown in Fig. 1.3. The causal (independent) variable acts on a situation, the nature of which is established by the parameters. The effect (dependent) variable is the product of the interaction between the causal variable and the situation. Alternatively, each block in the block diagram can be taken to represent a conveniently bounded sub-system, or bit of "machinery". The arrows indicate the way in which the various sub-systems are interconnected and summation of variables is indicated by the summing symbol illustrated in Fig. 1.2.

The main utility of block diagrams is that they provide a topological representation of a system in a manner that focuses attention on the parameters and sub-systems. At the same time, the system is described in a rigorous and unambiguous manner. Thus it is important, in using block diagrams, to stick to the conventional symbols, since these provide a powerful vehicle for accurate description and their use involves a disciplined attitude to the subject under discussion. In addition to providing a rigorous method of description, block diagrams introduce a new way of thinking about systems. This is the "black box" approach in which the interest lies in establishing the relationship between the inputs and outputs of a sub-system, rather than in the nature of the sub-system itself. In other words: it is the behaviour of variables which is of prime importance, rather than the nature of the mechanisms responsible for the behaviour.

1.3 TRANSFER FUNCTIONS

Any block in a block diagram can be considered as a black box and, in specifying the relationship between the input and output variables, it is necessary to describe the operation of the contents of the box on the input, to provide a given output. However, the operation may be such that the output may be changed in relation to time as well as in magnitude. The operation may delay the output, for example. Thus every block in a block diagram must be considered both in terms of the magnitudes of the variables involved and in terms of their behaviour as a function of time. This situation is

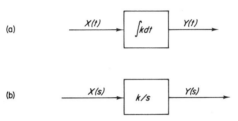

Fig. 1.4. Block diagram representation of a relationship between two variables: (a) as a function of time, (b) as a function of the Laplace operator s.

illustrated in Fig. 1.4a. The behaviour of the variables involved can be calculated by means of classical calculus, but it is more convenient to handle this type of problem by employing "operational calculus" and this is standard practice in control theory.

In operational calculus the differential coefficient d/dt can be replaced by the "differential operator" p. For example

$$\text{if } y(t) = \frac{dx}{dt}, \quad \text{then} \quad y(p) = xp. \tag{1.12}$$

The differential operator is useful for the more simple aspects of control theory, but control engineers generally employ a more powerful method, based on the "Laplace operator" s. The general procedure is to take a function of time, $f(t)$, and transform it into a function of the complex variable s, by finding the "Laplace transform" of the time-function. Thus $\mathcal{L}f(t) = F(s)$, which should be read as "the Laplace transform of the function of time is a function of s". The Laplace transform can be formally defined as follows

$$F(s) = \int_0^\infty f(t) \, e^{-st} \, dt \tag{1.13}$$

In practice, it is not usually necessary to work things out from first principles, as the Laplace transforms of most simple functions can be looked up in a table. Table 1.2 gives some of the more common functions. The usual procedure is to determine the Laplace transform of the necessary time-functions from the table; work out the problem in terms of s; and then look up the "inverse" Laplace transform, to obtain the answer as a function of time. This procedure is as simple as that for using logarithms.

Table 1.2. Some useful Laplace transform pairs

Time function	$f(t)$	Laplace transform $F(s)$
Unit impulse	$\delta(t)$	1
Unit step	$u(t)$	$1/s$
Unit ramp	t	$1/s^2$
Delayed impulse	$\delta(t-T)$	e^{-Ts}
Exponentials	e^{-at}	$\dfrac{1}{s+a}$
	$\dfrac{1}{a}(1-e^{-at})$	$\dfrac{1}{s(s+a)}$
Sine wave	$\sin \omega t$	$\dfrac{\omega}{s^2+\omega^2}$
Damped sine wave	$e^{-at}\sin \omega t$	$\dfrac{\omega}{(s+a)^2+\omega^2}$
	$\dfrac{1}{(s+a)(s+b)}$	$\dfrac{1}{b-a}(e^{-at}-e^{-bt})$
	$\dfrac{s}{(s+a)(s+b)}$	$\dfrac{1}{a-b}(ae^{-at}-be^{-bt})$

For most practical purposes, s can be considered as identical to p. Thus, rather than use both p and s, it would seem simpler to stick to s, even though this will often seem like using a sledge-hammer to crack a nut.

Returning to the subject of block diagrams, the Laplace equivalent of Fig. 1.4a is shown in Fig. 1.4b. When the input and output variables are described in terms of s, the operation by which one is obtained from the other is called the "transfer function" of the sub-system. The transfer function $= y(s)/x(s)$ for zero initial conditions (see section 3.2.1). A system is said to be "linear" if its transfer function is always the same, irrespective of the input. An important, and convenient, property of linear systems is that their transfer functions can be treated algebraically, thus greatly simplifying the mathematical procedures involved in the analysis of the behaviour of such systems. This facility provides the main rationale for the use of operational calculus in control theory.

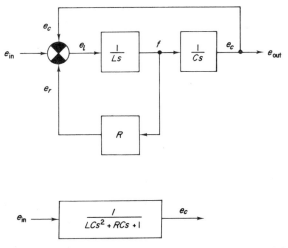

Fig. 1.5. Block diagram from Fig. 1.2 portrayed in operational form: (a) the synthesized system, (b) the overall transfer function.

As an example, let us return to the system illustrated in Fig. 1.2. From the block diagram of Fig. 1.5, which portrays the system in operational form, it can be seen that the transfer functions of the three elements are as follows

$$f/e_l = 1/Ls; \qquad e_r/f = R; \qquad e_c/f = 1/Cs,$$

thus in general, transfer function = output/input. From a knowledge of the transfer functions of the individual elements it is possible to obtain the overall transfer function of the system, e_c/e_{in}. Starting from the equation indicated by the summing point

$$e_l = e_{in} - e_r - e_c = e_{in} - \frac{1}{LCs^2}\, e_l - \frac{R}{Ls}\, e_l = e_{in} - e_l\left(\frac{1}{LCs^2} + \frac{R}{Ls}\right)$$

$$e_{in} = e_l + e_l\left(\frac{1}{LCs^2} + \frac{R}{Ls}\right) = e_l\left(1 + \frac{1}{LCs^2} + \frac{R}{Ls}\right)$$

$$e_l = \frac{e_{in}}{1 + \dfrac{1}{LCs^2} + \dfrac{R}{Ls}}, \quad \text{but} \quad e_c = \frac{e_l}{LCs^2},$$

thus

$$e_c = \frac{e_{in}/LCs^2}{1 + \dfrac{1}{LCs^2} + \dfrac{R}{Ls}} = \frac{e_{in}}{LCs^2 + RCs + 1}.$$

Hence the overall transfer function

$$e_c/e_{in} = \frac{1}{LCs^2 + RCs + 1}.$$

Although it is possible to calculate the overall transfer function for any completely specified block diagram, for complex systems the above method can become rather involved. A more convenient method can be achieved by the use of flow graphs.

1.4 FLOW GRAPHS

Flow graphs are formally identical to block diagrams and the difference between the two is one of graphic convention only. However, for some purposes, flow graphs have considerable advantages over block diagrams.

Block diagrams draw attention to the parameters and sub-systems, which are represented by "boxes". Flow graphs, in contrast, accentuate the variables of the system and represent summation in a more natural manner. The difference in the summing point is largely responsible for the superior flexibility of flow graphs.

In the notation used for flow graphs, variables are represented by small circles called "nodes", and are connected by lines called "branches", which carry a label indicating the functional relationship between the variables. This label consists of an arrowhead indicating the direction of causality and an associated "transmittance" representing the type of function involved. Flow graph transmittances are generally linear and can thus be manipulated algebraically, as shown in Fig. 1.6a. When a variable is dependent on two others or more, the branches converge on the dependent variable, as shown in Fig. 1.6b. Thus each node acts as a summing point in a linear system. Differences in sign are indicated by the transmittances.

The main advantage of flow graphs is their flexibility, which has been put to good use by engineers, in the formulation of methods of flow graph reduction. By flow graph reduction it is possible to short-cut much of the mathematical procedure involved in calculating transfer functions, or the behaviour of any chosen variable during system operation.

Most of the procedure for flow graph reduction follows from the two basic rules of construction: that the value of the dependent variable is obtained by multiplying the transmittance by the independent variable (Fig. 1.6a); and that branches entering a single

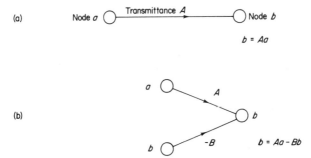

Fig. 1.6. Flow graph representation of linear relationships between variables.

node are summed (Fig. 1.6b). Nodes which have only one branch entering and one leaving, are called "cascade nodes". These can be eliminated, if desired, by multiplying the two transmittances (Fig. 1.7a). Conversely, it is often useful in analysing a system to insert temporary "dummy variables" into the graph. This can be done by interposing a node into a branch and putting a unit transmittance on one side (Fig. 1.7b). When two nodes are connected by parallel branches in the same direction the total effect can be represented in a single branch, by summing the transmittances (Fig. 1.7c). A single branch can be split into two by the same principle. Ingoing branches can be shifted by multiplying the outgoing transmittance by the ingoing transmittance (Fig. 1.7d). This procedure often results in the elimination of branches entering a closed path, leaving only a "self-loop", which is a node with a branch returning to itself (Fig. 1.7e). In a path containing a self-loop the path transmittance G is increased by the loop transmittance g, for each additional traverse of the loop. Hence

$$G = 1 + g + g^2 + g^3 + g^4 + \ldots .$$

The expression obtained from this series is

$$G = \frac{1}{1-g},$$

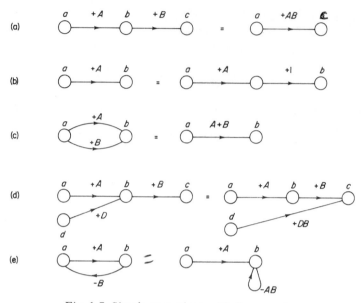

Fig. 1.7. Simple operations with flow graphs.

when the magnitude of $g < 1$, which is the basic equation of feedback theory.

As an example of the flow graph method the transfer function of the system portrayed in Fig. 1.5 is calculated by flow graph reduction. This procedure is illustrated, step by step, in Fig. 1.8, using the rules illustrated in Fig. 1.7. Essentially, this is a graphical procedure for solving simultaneous differential equations by means of operational calculus. In practice, it consists merely of a series of simple algebraic manipulations.

Transfer functions can also be calculated from flow graphs by direct methods. As these methods are useful only for very complex flow graphs, the reader is referred to more advanced treatments of the flow graph method (Chow and Cassignol, 1962; Lorens, 1964).

Since their invention by Mason (1953, 1956), the use of flow graphs has grown considerably in various branches of engineering and many advances have been made in their application to linear, non-linear, and statistical systems (e.g. Lorens, 1964; Huggins and Entwisle, 1968). However, block diagrams are still more widely used and for some purposes they are preferable. Both methods will be used in this book, mainly for describing systems in a precise and rigorous manner. Throughout the rest of the book it will be assumed

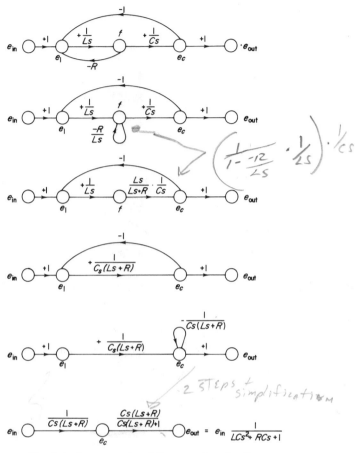

Fig. 1.8. An example of flow graph reduction.

that the reader is familiar with the basic language and symbolism of control theory. New concepts will be discussed as they are introduced, but as a number of texts on the application of control theory to biological systems are now generally available (e.g. Bayliss, 1966; Milhorn, 1966; Milsum, 1966), little more than an outline of the various aspects of control theory will be given.

1.5 SYSTEMS

A system may be defined as an ordered arrangement of physical or abstract entities; and a behavioural system can be envisaged as a network of functionally related variables, relevant to the behaviour in question. It is useful to distinguish between the "physical system" and the investigator's representation of this, which may be called an "abstract system".

In addition to the variables which make up a system there are variables outside the system, which can affect its behaviour. These external variables may be "control variables", which represent the stimuli to which the system is designed to respond, or they may be "disturbance variables", the effect of which the system is generally designed to minimize. External variables may be regarded as properties of the system environment, or they may belong to other systems to which the system under study is joined. The latter case is particularly important when the system under consideration is really a sub-system. Engineers are usually able to specify the external variables acting on the system under study, so that the relationship between the system and its environment presents no great difficulty. However, in biological systems this is not always the case.

Strictly speaking, the whole animal constitutes the system; and an abstract system, which represents a particular aspect of behaviour, is really a sub-system belonging to a very large and complex system. Thus, while the engineer can study systems of manageable proportions, the biologist is faced with a system of vast complexity and must try to isolate a sub-system that is suitable for study. So the main difference between physical and biological systems is that the external variables influencing the former are environmental, whilst those influencing the latter generally belong to other sub-systems. This difference is of great practical importance, because environmental variables are much easier to measure than variables which are internal to the system as a whole.

The complexity of biological systems makes selection inevitable and the degree of detail in which the investigator is interested will largely determine the size of the sub-system which he isolates for study. The behaviour of a single muscle can be studied at a detailed level, for example, whereas the behaviour of a whole limb must, initially, be analysed at a less detailed level. In this book it will be shown that large sub-systems can be simulated by relatively simple abstract systems, if the degree of detail is low. Parts of the simple abstract system can then be isolated for more detailed study, and this will generally lead to an increase in the complexity of the abstract system. Thus the biologist can choose either a relatively discrete sub-system and study it in detail or he can make an overall analysis of a large part of the system, and hope to obtain some insight into a method of analysing its sub-systems in greater detail. Both of these methods of approach have been used in the application of control theory to behaviour.

The main purpose of systems analysis is the accurate prediction of the behaviour of a machine under all possible types of influence by external variables. In order to describe the behaviour of the machine the investigator must measure certain essential variables. As measuring instruments are themselves machines, it is essential to ensure that the results represent the true behaviour of the system under study and are not contaminated by the behaviour of the "measuring system". In biological systems it is generally possible to measure only a few essential variables and the behaviour of other variables must be inferred from these direct measurements.

The first stage of systems analysis is to devise a "preliminary model" which is based on the data obtained from the measurements. The preliminary model consists of an abstract representation of the network of cause and effect, and is intended to mimic that of the physical system. As a preliminary model can lead only to qualitative predictions about the system behaviour, it is necessary to develop it into a "mathematical model" by specifying the numerical values of the parameters concerned. From a mathematical model it is possible to make quantitative predictions which can be compared with the observed behaviour of the physical system. After a comparison of observed and predicted behaviour it may be necessary to modify the measuring technique, the preliminary model, or the mathematical model. As systems analysis procedure is an iterative process it leads to a closer and closer agreement between the observed and predicted behaviour. When such agreement is reached an "abstract system" is attained and this should then be tested experimentally.

CHAPTER 2

Open- and Closed-loop Control

When the value of a dependent variable Y is determined uniquely by an independent variable X, the behaviour of Y may be said to be "controlled" by X. When the value of Y in no way influences the value of X, the control can be said to be "open-loop", in contrast to "closed-loop" control in which some such influence does occur. Diagrams of simple open- and closed-loop control configurations are illustrated in Fig. 2.1.

Open-loop control

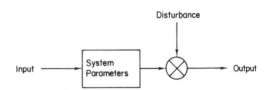

Closed-loop control

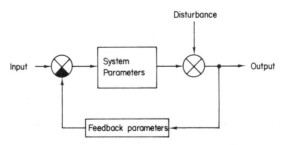

Fig. 2.1. Simple open- and closed-loop control configurations.

In the case of open-loop control, the value of the output is liable to be affected by disturbance variables affecting the output directly, or by changes in the parameters of the control mechanism. In a closed-loop control system, however, such disturbances can be compensated for by "feedback" from the output such that the input is appropriately adjusted. This is the main advantage of closed-loop control. Nevertheless, open-loop control can have certain advantages and examples are not difficult to find in the behaviour of animals.

2.1 OPEN-LOOP CONTROL

Open-loop control can be found in cases where disturbance of the output during normal functioning is extremely unlikely. For example, the eyeball is not normally a loaded organ and we might expect control of eye position to differ from that of the limbs, which are frequently subjected to mechanical loading and disturbance.

The present evidence suggests that eye movement control is achieved without proprioceptive feedback, although visual feedback is used during some types of visual tracking. Helmholtz (1867, 1962) proved that there is no position sense in the eye muscles. Using evidence from mechanical manipulation of the eyeball, the positions of after-images, and apparent motions induced by attempts to move the eye when the extraocular muscles are paralysed, Helmholtz concluded that "our judgments as to the direction of the visual axis are simply the result of the effort of will involved in trying to alter the adjustment of the eyes". Despite the contrary views of William James (1890) and Sherrington (1918), present-day opinion is in agreement with Helmholtz (Merton, 1964; Howard and Templeton, 1966).

As a matter of common observation, when the eyeball is displaced in its socket by finger pressure, the visual axis is shifted (as can be seen from the double image) and it remains shifted as long as the finger pressure is maintained. The eyeball does not push back against the finger in an attempt to regain its previous position. This failure to compensate for the disturbance is characteristic of open-loop control. In contrast, a closed-loop control system acts to counteract the disturbance. For example, the speed of a gramophone turntable is governed by a feedback system which is designed to take account of the drag of the needle when a record is played on it. If one attempts to brake the turntable with one's finger, the power is increased to compensate for the load and the turntable is felt to push back against the finger. Similarly, in the control of limb position, proprioceptive feedback serves to compensate for load.

Although muscle spindles' probably play some part in compensation for some types of disturbance, only joint receptors are responsible for the measurement of position *per se*. Thus, whereas the control of limb position can be said to be achieved with the aid of feedback from the joint receptors, the control of eye position is achieved by motor outflow only. The situation is summarized in Fig. 2.2. In the absence of load, motor output is sufficient for accurate control of eye position. Merton (1961) found that the gaze

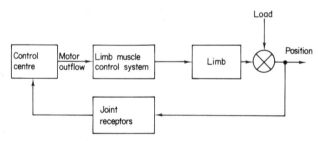

Fig. 2.2. Outline of eyeball and limb position control systems.

could be redirected in the dark towards an object previously fixated in the light. The accuracy of fixation in the dark was fairly good, but much less good than the accuracy in the light, when visual feedback is additionally available.

An important property of open-loop control is that it permits a much more rapid change of output than is possible with closed-loop control. Open-loop control is found in rapid eye and limb movements in humans and in rapid prey-catching movements in insects and cephalopods, for example.

Messenger (1968) has shown that in the cuttlefish *Sepia officinalis*, which catch small crustaceans by means of two extensible tentacles

(Fig. 2.3), prey seizure is essentially an open-loop response. The control of attack falls into two phases. The first is essentially a visually guided closed-loop system, in which the prey is fixated binocularly and movement of the prey is followed by movement of the cuttlefish, such that the visual error is reduced almost to zero. Only when this "dead reckoning" phase is complete are the tentacles ejected. The final seizure phase is so rapid (about 30 ms) that there can be no time for the tentacles to be visually guided onto the prey. The prey seizure is essentially an open-loop response, and this conclusion is verified by the finding that prey capture is not affected

Fig. 2.3. Cuttlefish (*Sepia officinalis*) catching a shrimp. (After Wells, 1962.)

if the light is turned off at the moment of tentacle ejection. In addition, the prey is missed if it is moved during ejection. In other words, the system is unable to compensate if the position of the target is disturbed during the final phase of attack. A similar system is found in the mantids investigated extensively by Mittelstaedt (1957) and this work will be discussed later in the chapter.

In a complex system, open-loop control is likely to exist only with respect to certain restricted aspects of the overall control system. In the case of eye movements, for example, proprioceptive control may be open-loop but closed-loop control is also available in the form of visual feedback. The phenomenon of thumping in rabbits may be used to illustrate this principle at a different level. Charles Darwin (1872) observed: "Rabbits stamp loudly on the ground as a signal to their comrades." The evidence for the function of this behaviour is largely anecdotal, but Black and Vanderwolf (1969) have demonstrated that thumping behaviour can be elicited by electrical stimulation of the brain and by peripheral electric shock. Their evidence suggests that thumping is a sign of fear. The control of this behaviour, in the short-term sense, probably involves proprioceptive feedback loops. However, it is also important to consider the more long-term consequences of the behaviour. Many behaviour patterns have consequences which feed back and affect the performance of the activity, via interaction with the environment. Feeding behaviour, for example, can alter the environment in a material sense; while courting behaviour may alter the behaviour of other individuals. The closed loop that is formed by the animal's reactions to these self-produced environmental changes may have important implications for the control of ongoing behaviour. In the case of thumping, however, there is no evidence for such a feedback loop. Although thumping may affect the behaviour of other individuals this will be of no consequence to the thumper. In this sense, therefore, thumping may be regarded as open-loop behaviour. The point of this illustration is that, in using the terms open-loop and closed-loop control, the level of description should be specified as, at different levels, different types of control may exist within the same system.

In summary, it may be said that the main advantage of open-loop control is speed of response, while the main disadvantage is its inability to cope with disturbances to the output. In practice both open- and closed-loop control are likely to coexist, although they may operate at different levels.

2.2 CLOSED-LOOP CONTROL

In general, closed-loop systems are characterized by the fact that the value of dependent variables can influence the value of "independent" variables or parameters. The mechanism by which such influence is achieved is usually called a "feedback mechanism". From a mathematical viewpoint, feedback exists between two variables whenever each affects the other. For example, let W_{in} be the rate of inflow of water into a bath, and W_{out} the rate of outflow. The amount of water in the bath at any time is equal to the initial amount in the bath, plus the time integral of the net rate of inflow. Suppose that small changes in the rate of outflow are proportional to the amount in the bath, outflow being greater when the water pressure is higher

$$W_{out} = K \int_0^t (W_{in} - W_{out})dt. \qquad (2.1)$$

The situation can be expressed in terms of a simple block diagram (Fig. 2.4), and from this it appears that feedback exists between outflow and inflow. There is no feedback in any physical sense, but because feedback exists in the mathematical formulation a "fictitious" feedback loop must be introduced into the block diagram. Physical feedback should always be represented by a "feedback parameter" in the block diagram, while fictitious feedback, the value of which is always unity, is portrayed without any such parameter as in Fig. 2.4. In Fig. 2.1, on the other hand, the feedback is physically meaningful and the feedback parameters are represented by the appropriate block in the block diagram.

The physical feedback mechanism usually consists of a "sensor", a measuring device which monitors the output variable; and of a mechanism to convey the information to the control centre. For

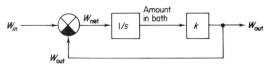

Fig. 2.4. Fictitious feedback, illustrated by a simple hydraulic system.

example, in Fig. 2.2 the limb position is monitored by the joint receptors and the information is fed back to the CNS via the appropriate nerves, which are not shown in the diagram, but which strictly constitute a part of the feedback mechanism.

"Positive feedback" exists when the consequences of the monitored output tend to accentuate further output. In a linear system the feedback is added to the input and the controlled device is actuated to an ever-increasing extent. Positive feedback systems are inherently unstable, since action and reaction intensify one another forming a vicious circle. Nevertheless, positive feedback systems do exist as behavioural phenomena, the population explosion being a good example. Moreover, when incorporated into complex systems with built-in checks, positive feedback loops can have certain advantages.

When the consequences of the monitored output of a system tend to diminish further output "negative feedback" is said to exist. In a linear negative feedback system the output of the feedback mechanism is subtracted from the system input to give an "error" variable, as illustrated in Fig. 2.5. The error actuates the controlled

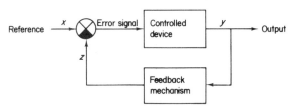

Fig. 2.5. Error-actuated negative feedback system.

device, so that the output tends in a direction opposite to that which caused the error. Thus a large positive error would result if the output were much smaller than the input and this would actuate the controlled device so as to increase the output. The output of the feedback mechanism would then increase and the error would be consequently reduced. Thus it can be seen that a negative feedback system always tends to minimize the error and such systems tend towards an equilibrium.

The equilibrium point of a negative feedback system can be achieved through "passive" or "active" control (Milsum, 1966). Although this distinction is extremely difficult to formalize it has important implications. Ashby (1960) points out that "Every stable system has the property that if displaced from a state of equilibrium and released, the subsequent movement is so matched to the initial displacement that the system is brought back to the state of equilibrium." However, it is useful to distinguish between systems that attain equilibrium as a result of the cancelling out of forces within the system and systems in which the equilibrium point is related to an explicit reference value imposed from without.

As an example, consider the systems outlined in Figs 2.4 and 2.5. In the latter, the device is controlled by the "error" variable, $x - z$, This error is the result of a comparison between the input and the consequences of the output that are relayed by the feedback mechanism. It is conceivable that the input could be a "reference" signal established as a standard of comparison for the feedback control system and a machine called a "comparator" could carry out this function. Many household electric fires, for instance, are provided with a control dial by means of which a "desired" temperature can be set. This dial forms part of the thermostat, which provides the necessary negative feedback, and turns the fire off when the desired room temperature is attained. In the case of the system outlined in Fig. 2.4, it is inconceivable that the system is responding to an externally imposed reference signal, even though the net inflow, W_{net}, is formally identical with the error variable in Fig. 2.5. The equilibrium reached by the bath water is entirely the result of forces acting within the system. The level of water in the bath can be said to be under "passive" control, in contrast to the "active" control exercised by a thermostat.

The distinction between passive and active control hinges around the question whether a comparator is, in principle, physically identifiable within the system or whether it is possible to identify a reference value or "set-point", the value of which is independent of the system under consideration. The traditional concept of temperature regulation in physiology is that the system operates as a balanced dynamic system, but recent evidence suggests that physiological control of body temperature may be achieved through a set-point regulator. Hardy (1965), after reviewing the relevant control equations, concludes that "the system performs 'as if' it were a set-point system, but it could also be a combination of two

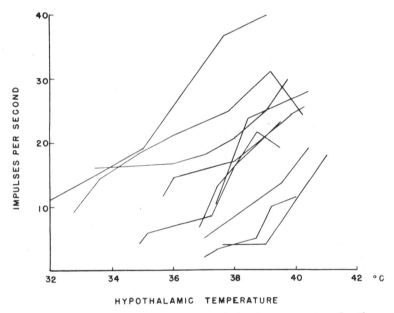

Fig. 2.6. Impulse frequency versus hypothalamic temperature for thermally responsive neurones. (After Hardy, 1965.)

balanced, dynamic systems with proportional control only". In order to settle the question, it would be necessary to identify the mechanism which provides the set-point. In practice this amounts to the identification of neurones within the CNS, which are insensitive to local temperature changes. The existence of such neurones in the anterior hypothalamus was demonstrated by Nakayama *et al.* (1963). These workers found that the discharge rate of single neurones in the anterior hypothalamus increased as a result of local warming, achieved by circulating water through implanted stainless steel tubes (Fig. 2.6). The impulse frequency of other neurones was found to increase with decreasing temperature and many neurones in the preoptic area were found to be insensitive to local temperature changes (Fig. 2.7). It might be that some of these neurones serve as the set-point for temperature regulation, but the function of these neurones is as yet unproved, although the preoptic area is known to be intimately involved in temperature regulation.

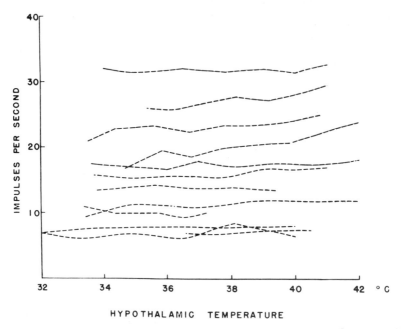

Fig. 2.7. Impulse frequency versus hypothalamic temperature for thermally unresponsive units of the preoptic region. (After Hardy, 1965.)

The concept of a set-point does not necessarily imply that the reference value is fixed for ever. It is probable that shifts in the set-point for temperature regulation do occur as a result of non-thermal influences such as sleep, exercise and fever. Hammel (1965) suggests that "the regulation of body temperature may be described as if the hypothalamus were responsive to changes in its own temperature and as if the set-point temperature for each response were adjusted by the environmental temperature, i.e. by its effect upon the skin temperature". Hammel proposes a neural model by which such a control system could be achieved.

The role of muscle spindles in the feedback control of limb movements can be taken as another example (Fig. 2.8). The sensory portion of the spindle does not respond to the absolute length of the

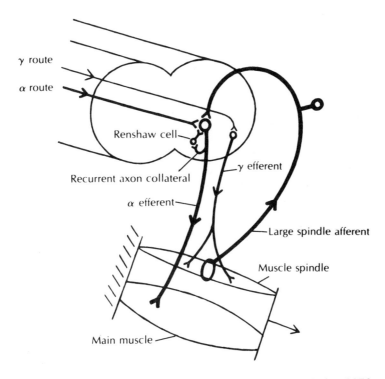

Fig. 2.8. Innervation of the muscle spindle. (From Hinde, 1970, after Hammond *et al.*, 1956.)

muscle but, as Hinde (1970) points out, "to the difference between the length of the spindle and the length of the muscle. Whether the muscle contracts or relaxes depends in part on the state of contraction of the spindle". Hinde goes on to say that "we shall have frequent occasion to refer to feedback control of this sort, and it will be convenient to have a general term for the required or optimal stimulation received when the effector organ is in its equilibrium position. A precise English word is hard to find, though such terms as 'goal', 'target value', or 'equilibrium position' are all useful in some contexts. The German word 'Sollwert' (literally the 'should be value') has a wider applicability. In the present case the Sollwert is achieved when the lengths of muscle and muscle spindle bear a certain relation to each other, but the term might equally be applied to the setting on the thermostat controlling a heating system, or that on the governor controlling a steam engine". The problem here is that the term Sollwert is used to equate the dynamic biasing of the spindles with the static set-point of a thermostat. The difficulty lies, not necessarily in the use of the term Sollwert in this particular context, but in the fact that the same term is also used for a "Schema", or "neuronal model" with which an external situation can be compared, as in the case of bird song (Hinde, 1970), and for the "efference copy" involved in the reafference principle of von Holst and Mittelstaedt (1950). As these are very different concepts, it would be better to distinguish between them and an attempt will be made to do so. But at this point it will be useful to define the terms by means of which feedback control systems can be unambiguously described and discussed.

An "input variable" is a variable which affects the behaviour of a control system, without itself being affected by any consequences of such behaviour. The boundaries of a system, or sub-system, can be defined with reference to the complete set of input variables acting upon it.

A "system variable" is a variable, within a particular system, the value of which is affected by input variables, or by other system variables.

An "output variable" is a variable, arbitrarily selected for measurement by an observer and capable of giving useful information about the behaviour of the system.

A "mixing point" is a device whose output is equal to some function of two or more inputs.

A "command" is a variable, or set of instructions, established by some means external to, and independent of, the system under consideration.

A "reference variable", or "set-point", is an input variable, or a command, established as a standard of comparison for a feedback control system.

A "disturbance variable" is an input variable, which causes the value of an output variable to deviate from the level set by a reference variable, or command.

A "comparator" is a mixing point whose output is equal to the sum, or difference, between two or more variables.

A "feedback variable" is a function of an output variable, which affects other system variables in such a manner that a closed-loop is formed.

An "actuating variable" is a system variable which serves as the input to a particular sub-system.

An "error variable" is the output of a comparator.

An "error-actuated" sub-system is a sub-system which is actuated by the output of a comparison between a reference variable and a feedback variable.

It should be noted that a reference variable is a type of command but that other types of command can also exist. A command can operate through a mixing point other than a comparator, or by directly altering the value of a parameter. Only commands that operate through comparators are called reference variables. The term Sollwert has been used both for reference variables and for other types of command, but the terms Sollwert and command are not synonymous. For example, the term Sollwert has been used for the "expected value", or "efference copy", in a reafference system (see section 7.1.2). The term command would not be used in this context. Thus, in the case of vision, the outcome of the comparison between the efference copy and the reafferent message can alter the nature of the perception without affecting the performance of the control system. In no sense is any performance criterion "commanded", nor need the efference copy act as an input variable to the eye-movement system.

At this point it may be helpful to consider a particular example in some detail. Mittelstaedt's analysis of prey capture in mantids illustrates a number of topics that have been discussed so far. In particular, both open-loop and closed-loop control are involved and block diagram and flow graph techniques can be profitably employed in their analysis.

2.3 PREY CAPTURE IN MANTIDS

Mantids are carnivorous insects which lie in wait throughout the day and catch flies which come within reach. The prey is detected by well developed compound eyes and is faced by movement of the head. The mantid then strikes at the prey with a very rapid movement of the forelegs. The short duration of the strike (10-30 ms) makes it unlikely that the forelegs are guided by visual feedback. The alternative is that the foreleg movement is ballistic and that the strike is aimed by some "dead reckoning" system.

The method of prey capture has been extensively studied by Mittelstaedt (1957), mainly in *Parastagmatoptera unipunctata* (Fig. 2.9). Mittelstaedt's (1957) study is concerned with the way in which the fully developed adult mantid aims its stroke in the horizontal plane. The direction of the stroke must be determined by information about the direction of the prey relative to the

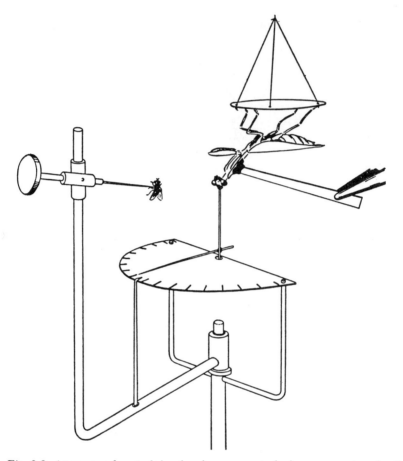

Fig. 2.9. Apparatus for studying head movements during prey-capture by the mantid (*Parastagmatoptera unipunctata*). (After Mittelstaedt, 1957.)

prothorax, to which the forelegs are attached. Consequently the animal must take into account not only the angle between the target and the centre of the head, but also the angle of the head to the prothorax. The angle of the target to the head must be detected by the compound eyes, and that of the head to the body by the hair plates near the neck which act as proprioceptors (Fig. 2.10).

When the mantid detects a prey it turns its head towards it. Because of this, Mittelstaedt argues, there must be a functional unit which transforms the position of the prey relative to the compound eyes into information which is used to direct the turning of the head. But as soon as the head changes its position, the position of the eye

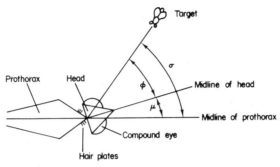

Fig. 2.10. Diagram of mantid-prey relations. (After Mittelstaedt, 1957.)

relative to the target is changed too. Thus the output of the head moving mechanism must influence the input to this mechanism as shown in Fig. 2.11.

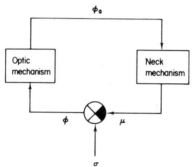

Fig. 2.11. Block diagram of closed-loop relation between optic and neck mechanisms.

The optic input ϕ equals the angle between the target and the head (Fig. 2.10) and the optic output ϕ_c is a (nervous) representation of this angle, which acts as the input to the neck mechanism. The output of the neck mechanism μ equals the angle between the head and the prothorax (Fig. 2.10). The angle between the target and the mid-line of the prothorax $\sigma = \phi + \mu$, therefore the optic input $\phi = \sigma - \mu$ and is diminished by any increase in μ. If the relationship between the system variables is linear, the optic mechanism and the neck mechanism can be represented by the parameters Q and N respectively. These parameters represent the "amplification factors" (Mittelstaedt, 1957) of the two mechanisms. Thus $\phi_c = Q\phi$ and $\mu = N\phi_c$. Having defined the system parameters in this way, the

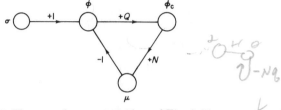

Fig. 2.12. Flow graph representation of Fig. 2.11.

ultimate value of the optic input ϕ can easily be found by the flow graph method. Figure 2.12 is a flow graph of the system in Fig. 2.11. By flow graph reduction,

$$\phi = \frac{\sigma}{1+NQ}, \quad \text{therefore} \quad \frac{\phi}{\sigma} = \frac{1}{1+NQ}. \tag{2.2}$$

Mittelstaedt called ϕ/σ the "fixation deficit" because it corresponds to the smallest value of ϕ relative to σ that the animal is able to achieve with a given value of the "system amplification factor" NQ. For example, if $NQ = 4$ and $\sigma = 20°$, the final value of

$$\phi = \frac{\sigma}{1+NQ} = \frac{20}{1+4} = 4°, \tag{2.3}$$

and the fixation deficit

$$= \frac{\phi}{\sigma} = \frac{4}{20} = \frac{1}{5}. \tag{2.4}$$

Thus, whatever the initial values of ϕ and σ, the final value of ϕ will always equal $\sigma/5$, for the given amplification factor $NQ = 4$.

As the mantid turns its head towards the target, the hair plates will be activated. Thus the neck mechanism output acts as a proprioceptive input (Fig. 2.13). Mittelstaedt proposes that the proprioceptive output δ_c is subtracted from the optic output ϕ_c and that the difference μ_c becomes the neck mechanism input, as shown in Fig. 2.13. The effect of this additional feedback mechanism is to enable the neck mechanism to compensate for extraneous disturbances or additional loads, such as might occur if the mantid already had a fly in its mouth.

Suppose that the direction of the stroke κ is defined as the angle between the endpoint of the stroke and the mid-line of the prothorax; then the prey will be hit if $\kappa = \sigma$. From the system outlined in Fig. 2.13 it is easy to calculate that the fixation deficit

$$= \frac{\phi}{\sigma} = \frac{1+PN}{1+N(P+Q)} = k. \tag{2.5}$$

Thus

$$\sigma = \frac{\phi}{k} = \kappa,$$ (2.6)

so that the mantid can always obtain the correct strike angle from the optic input ϕ, when the system has stopped moving. As the optic

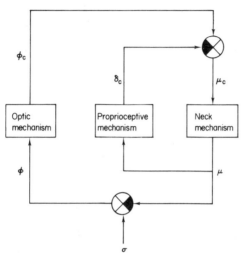

Fig. 2.13. Incorporation of the proprioceptive feedback loop into the system for head-position control.

input cannot be used without being first measured, Mittelstaedt proposed that the strike mechanism is actuated by the optic output $\phi_c = Q\phi$, as illustrated in Fig. 2.14, in which the amplification factor of the strike mechanism

$$= \frac{1}{Qk} = \frac{1 + N(P+Q)}{Q(1 + PN)}.$$ (2.7)

In other words, according to this hypothesis, the output of the dead reckoning system at steady state, which equals ϕ_c, is multiplied by a constant factor to give the correct stroke angle κ. Mittelstaedt then tested this hypothesis experimentally, and found:

(1) That elimination of the hair plates by nerve section (bilateral total deafferentation) causes the mantid to miss the prey on 70-80% of trials compared with 10-15% in the intact animal. The mantid usually strikes at too small an angle, and thus has a good chance of hitting the prey only when it is in line with the prothorax mid-line. According to the hypothesis (Fig. 2.14), the effect of the operation

is to eliminate the proprioceptive feedback loop. The fixation deficit will consequently be reduced from

$$k = \frac{1 + PN}{1 + N(P + Q)} \tag{2.8}$$

to

$$k' = \frac{1}{1 + NQ}. \tag{2.9}$$

Assuming that the operation does not change any parameters in the strike mechanism, the strike angle κ will be smaller than normal because of the reduction in the optic output ϕ_c. Thus the mantid will usually miss the prey except when it is in line with the body axis.

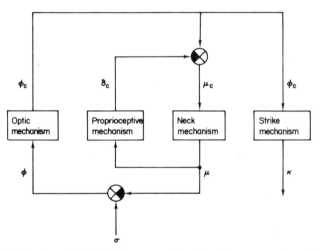

Fig. 2.14. Mittelstaedt's (1957) proposed system for prey-capture in mantids. σ = angle between prey and mid-line of prothorax, ϕ = optic input angle, ϕ_c = nervous representation of optic input, μ = angle between head and prothorax, μ_c = nervous input to neck mechanism, ζ_c = output of proprioceptive mechanism, κ = strike angle.

(2) When the relative position of the head and prothorax is fixed by means of a balsa wood strut, which does not touch the neck region, the animal aims with a bias opposite to that of the head deviation and at an angle proportional to the angle of head deviation. When the head is fixed on the mid-line of the prothorax there is zero bias. Fixation of the head relative to the prothorax blocks both the optic and proprioceptive feedback loops. The optic output ϕ_c will

now be determined by the angle of the head to the prothorax, so that

$$\phi_c = Q\phi = Q(\sigma \pm \mu),$$ (2.10)

the sign of μ being determined by the direction of head tilt with respect to the target (Fig. 2.10).

(3) Unilateral deafferentation of the hair plates combined with head fixing indicates that the effect of the operation is additively superimposed on the effect of head fixing. Thus a bias caused by head fixing is enhanced by operation on the side to which the head is deflected and counteracted by operation on the opposite side. These results are clearly contrary to those expected on the basis of the present hypothesis, since the proprioceptive loop should have no effect at all when the head is fixed. The results suggest that the strike mechanism is actuated by the optic output ϕ_c plus some function of the proprioceptive output δ_c.

(4) When the head of the intact animal is loaded with an extraneous mechanical force, which increases the angular momentum of the head not more than 50 mg/cm for a head weighing 25 mg, there is no significant decrease in aiming performance. This result is in agreement with the hypothesis, since the proprioceptive loop should be able to compensate for any extraneous disturbances to the neck mechanism and for changes in the normal state of the neck muscles.

The results of these experiments provide qualitative support for Mittelstaedt's hypothesis and suggest that the proprioceptive output also plays some part in actuating the strike mechanism. Quantitative evidence comes from experiments in which Mittelstaedt fixed the mantid by the prothorax and presented flies at a measurable direction and measured the position of the head μ, after the mantid had faced the fly. Two methods of measurement were used, one of which is illustrated in Fig. 2.9. Mittelstaedt discovered that there are two sorts of head movements; quick saccadic head movements occur when the target is more than 30 mm from the mantid, twice the distance of the stroke. Smooth continuous movements occur when the fly is within 30 mm. Obviously the system is more complicated than that described above. However, Mittelstaedt obtained a fixation deficit of 15% for continuous movements which was reduced after total proprioceptive deafferentation. From the fixation deficit it can be calculated that the overall gain of the system is about six. Mittelstaedt's work to this date (1957) provides a sound working hypothesis about the dead reckoning system by which the mantid aims its strike. The hypothesis is a good example of the method by which an abstract system can be built up logically from first principles and then tested experimentally. Inevitably the experimental work shows that the system is not adequate in all respects, but the results point the way to further research. It should be noted that, by deliberately ignoring the dynamic behaviour of the system, Mittelstaedt is able to construct an abstract system which is perfectly adequate for the purpose of erecting a working hypothesis.

CHAPTER 3

Behaviour as a Function of Time

Although it is a truism to say that behaviour is a function of time, this aspect of behaviour is by no means the only, or even the most common, focal point for behaviour study. In studying genetical or evolutionary aspects of behaviour, for instance, the way in which behaviour changes from one generation to the next is of prime importance. Similarly, in learning studies behaviour is generally considered as a function of trials, or some other measure of experience. In studying behaviour as a function of time one is primarily concerned with the performance of particular behaviour patterns, their orientation, goal directiveness and motivation. The manner in which such behaviour is controlled by internal and environmental stimuli and control theory is particularly appropriate to this type of study.

The degree to which control theory can be usefully applied to behaviour depends largely upon the nature of the behavioural measures employed in the study. As the type of measurements puts a number of constraints upon the validity of the method of analysis, it is appropriate to discuss the measurement of behaviour in some detail.

3.1 MEASUREMENT OF BEHAVIOUR

3.1.1 Levels of Measurement

Stevens (1951) points out that to perform operations with numbers that have been assigned to observations, there must be some isomorphism between the empirical observations, and the numerical system employed in their measurement. To be isomorphic, the structure of two systems must be the same in the mathematical operations they allow. The operations allowable on a given set of scores are dependent on the level of measurement achieved. Stevens discusses four levels of measurement and the operations that are permitted at each level.

(a) *The Nominal or Classificatory Scale*

In classifying phenomena, numbers can be used to identify the groups to which the phenomena belong. Such numbers constitute a nominal or classificatory scale, if the groups are mutually exclusive. For example, mutually exclusive aspects of feeding behaviour could be numbered in a purely arbitrary manner. Here the scaling operation is partitioning a given class into a set of mutually exclusive sub-classes. The members of any one sub-class must be considered as "equivalent" in the property being scaled. Thus if sniffing is taken to be diagnostic of a given sub-class, all occasions of sniffing must be relegated to that sub-class and all occasions must be considered as equivalent. The only kind of admissible descriptive statistics, at this level of measurement, are those, such as the mode, frequency count, etc., which would be unchanged by a transformation of the symbols which designate the various sub-groups.

(b) *The Ordinal or Ranking Scale*

An ordinal or ranking scale is achieved when a comparative relationship exists between mutually exclusive categories. For example, if category A is greater than, preferred to, or more desirable than, category B, then it can be stated that $A > B$. The properties of an ordinal scale are not isomorphic to the numerical system known as arithmetic, because the magnitude of differences between classes need not be equal. Great care must therefore be taken in applying statistics to this type of data. Appropriate statistical tests are discussed by Siegel (1956).

(c) *The Interval Scale*

When, in addition to there being a comparative relationship between mutually exclusive classes, the magnitude of the difference between classes is equal, interval scaling is achieved. An interval scale is characterized by a common and constant unit of measurement. Thus numbers may be associated with positions on the scale and operations of arithmetic may be meaningfully performed on the "differences" between these numbers. The ratio of any two such differences is independent of the unit of measurement. Thus the unit of measurement on an interval scale is arbitrary and the scale does not have a true zero. The Fahrenheit and Centigrade scales for measuring temperature are examples of interval scales.

(d) *The Ratio Scale*

The ratio scale has all the characteristics of an interval scale and in addition has a true zero point, so that the ratio of any two scale points is independent of the unit of measurement. Thus the ratio scale is entirely isomorphic with the structure of arithmetic and constitutes the most powerful level of measurement.

Having outlined the four levels of measurement discussed by Stevens (1951) and Siegel (1956) we can now relate these to the problem of measuring behaviour. All four levels of measurement depend on the existence of mutually exclusive classes and the first task is therefore to classify behaviour patterns in a suitable manner.

3.1.2 Classification of Behaviour Patterns

The most obvious mutually exclusive classes of behaviour are those which are logically incompatible. No animal can simultaneously stand up and sit down, or move forward and backward. However, there are other types of behaviour which can be seen to be incompatible only through knowledge of the muscles and reflexes involved. Thus Sherrington (1906) pointed out that pairs of reflexes which have some muscular movements in common are nevertheless neurologically incompatible, so that stimulation of one inhibits continued performance of the other.

In considering behaviour as a function of time the degree of incompatibility depends on the unit of time measurement employed. Thus if time were measured in whole seconds the animal could perform two neurologically incompatible movements within one time-unit. Suppose an observer were recording walking and pecking in a pigeon. He might record each peck by writing the symbol *P* with reference to the time scale, and each step by the symbol *W*. Using a one second time-unit his data might appear as follows,

Time-interval   1   2   3   4

Behaviour   *PPPPP PWW   WPW   WWW*

in which case walking and pecking would not appear to be incompatible, whilst on a 1/10 s time-unit they probably would. The decision to count *P* and *W* as mutually exclusive classes is largely an arbitrary matter, and the decision need not rest on the choice of time-unit. Thus the observer could decide *a priori* to regard *P* and *W* as mutually exclusive and could measure the time spent in each category, in which case the above data might be recorded as follows,

Occurrence	1	2	3	4
Behaviour	P	W	P	W
Time spent in seconds	1.2	1.0	0.2	1.4

An alternative method is to devise a set of mutually exclusive classes of behaviour and incorporate criteria by which one class can override another. Thus the observer could record W when the pigeon was walking, but not pecking, and P when the pigeon was pecking, whether it was walking or not. In this case the class P is defined by the occurrence of a peck within the time-unit employed, and this class overrides class W, so that the data would be recorded as follows,

Time-interval	1	2	3	4
Behaviour	P	P	P	W

Whatever measurement criterion the observer selects, he must ensure that the behaviour categories employed are mutually exclusive and he must be prepared to treat all members of a given category as equivalent. A nominal level of measurement will then have been achieved and certain data manipulations will be admissible. However, to measure behaviour as a function of time it is necessary to employ more powerful levels of measurement and this is not always easy. A ratio scale can be achieved by counting the number of events per unit time. The problem here is to be sure that successive events are equivalent. For example, the number of times a bird pecks at grain may be thought to be a measure of fairly discreet, and equivalent, events. But the pecks may differ in intensity or force and may vary in success in obtaining grain. Even the bar-presses of a rat in a Skinner box cannot be considered as absolutely equivalent, as they may vary in force or duration. The only satisfactory answer to this problem is that the proof of the pudding is in the eating. The fact that behavioural measures are not perfect means that there will inevitably be variation in behavioural data. A more serious consequence is that the scale of measurement may be non-linear with respect to the independent variable. Wherever interval or ratio scales are employed it is wise to conduct a test for linearity over the full range of scale of measurement used. Let us consider a rather extreme case.

In order to measure feather position in the Barbary dove (*Streptopelia risoria*), McFarland and Baher (1968) divided the body surface into six areas, as shown in Fig. 3.1. The delineation of the areas was essentially arbitrary, although casual observation suggested that these areas behave as functional units under certain circumstances. Three categories of feather position were employed: the feathers were scored as being "sleeked" (S) when they were obviously flattened against the body surface, and "raised"(R) when they were obviously erected above the normal. When they were in neither of these states, the feathers were scored as being "normal" (N). Each of these categories was given a score, so that $S = 0$,

Fig. 3.1. Body regions used in calculating the feather index. b = breast, c = crown, d = dorsal, n = neck, v = ventral, w = wing. The tail (t) was not scored. (From McFarland and Baher, 1968.)

$N = 1, R = 2$. During observations the feather posture predominating in each body region was recorded once per minute. In transcribing these observations the scores corresponding to the posture recorded for each region were summed within each 1-min observation period, to give a feather index (FI) with a minimum of 0 (fully sleeked), and a maximum of 12 (fully raised). Thus a type of scalogram technique was used to combine the scores from different body regions. This procedure is to some extent justified *post hoc* by the fact that no significant differences are found between the responses of the body regions to changes in ambient temperature, although there are variations from one observation period to the next.

On the face of it, only an ordinal scale is achieved by this method of measurement. There is no *a priori* means of knowing whether the intervals between categories are equivalent or not. Thus the numerical labelling, $S = 0, N = 1, R = 2$, is not strictly permissible. The same objection applies to the procedure of summing the scores from the different regions, since the areas of the body surface are arbitrarily chosen and are not of equivalent size. In practice, this method produces a feather index which has a fairly linear relationship with ambient temperature over the range 7-33°C (Fig. 3.2). Thus an ordinal scale appears justified retrospectively.

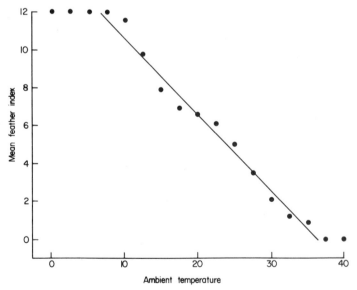

Fig. 3.2. Steady-state feather index as a function of ambient temperature. (After McFarland and Baher, 1968.)

3.1.3 Choice of Time-scale

In measuring behaviour as a function of time it is necessary to choose a suitable time-unit with respect to which the time spent at each activity is measured. The lower limit of the time-unit should be determined by the time taken to measure a change in the behavioural variable. For example, if it takes an observer a tenth of a second to record a behaviour item, he should not attempt to record items occurring faster than ten a second. Similarly, the response time of a measuring instrument should be taken into account in choice of time scale.

Although the maximum amount of information is obtained by employing a time-unit near to the minimum it will often be more convenient to choose a larger unit. Thus, as outlined below, larger time-units can be used to mask variability and rhythmicity in the data and also as a method of linearization. Figure 3.3 shows records of operant drinking behaviour in a dove. Rate of intake is plotted against time, using time-units of 1, 5 and 10 min. The same data are plotted in each case. When the data are scored as number of responses occurring within each 10 min period, the decline in performance appears to be exponential (a); but when a 5 min time-unit is used, it can be seen that there is an initial "warm-up" period, prior to the maximum response rate (b). Thus some important information is lost by employing too large a time-unit. With a 1 min time-unit, it can be seen that the decline in performance is due to the interpolation of pauses between bouts of responding at near the maximum rate (c). This phenomenon is characteristic of satiation of operant responding in doves (McFarland, 1970b). This degree of detail is not always necessary or even the most convenient method of recording the data. The record based upon the 1-min time-unit shows up the non-linearities in the behaviour which are induced by the all-or-none nature of the response rate as a function of time. McFarland and McFarland (1968) used a 5-min time-unit in their analysis of satiation of operant drinking behaviour in doves, in order to linearize the data and thus facilitate the application of classical control theory. This type of analysis leads to a description of the basic processes underlying the behaviour and opens the way for more detailed subsequent analysis.

Frequently, the high resolution achieved by the use of a small time-unit serves only to emphasize the variability of the data. In such cases a larger time-unit can often be employed in the interests of

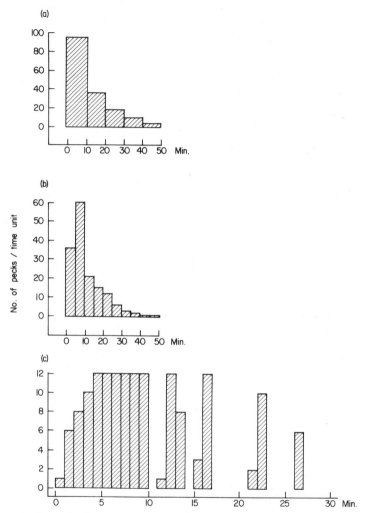

Fig. 3.3. Record of operant drinking behaviour in a dove. Rate of intake is plotted against time, using time-units of 10 (a), 5 (b), and 1 (c), min. (Unpublished observations.)

efficiency, without loss of valuable information. It can also be profitable to employ a larger time-unit for the analysis of rhythmical data. Many aspects of behaviour are subject to circadian rhythmicity which in some cases is superimposed on a more basic process. The body temperature of birds, for example, often shows cyclic variation of this type and in the analysis of long-term trends, such as that imposed by food deprivation, it is important to take measurements at the same time each day.

Finally, manipulation of the time-unit used in measurement can be used as a deliberate strategy in systems analysis. However, before exploring this possibility, it is necessary to consider some further aspects of behaviour as a function of time.

3.2 SYSTEMS ANALYSIS AS A FUNCTION OF TIME

There are three main approaches to the description of systems. A system may be described as a function of time, as a function of frequency, or in terms of state-functions. All three approaches will be discussed. Conceptually, description as a function of time is the most simple, though sometimes a less powerful, approach. In systems analysis, a common approach is to determine the response of a particular output to manipulation of a particular input. Three main classes of input function have been found to be suitable for such analytical operations. These are (1) transient inputs, (2) periodic inputs, (3) stochastic inputs. Transient inputs are particularly suited to analysis in terms of time functions, often called "transient analysis"; and periodic inputs to analysis as a function of frequency, called "frequency analysis". Stochastic inputs are related to more sophisticated methods of analysis in terms of time or frequency functions. For the present purposes we may restrict our considerations to the description and analysis of systems as a function of time. The classical approach to this type of problem is to consider the behaviour of the system in terms of differential equations.

3.2.1 Differential Equations

The input x and output y of a linear system are related by a differential equation of the form

$$a_0 x + a_1 \frac{dx}{dt} + a_2 \frac{d^2 x}{dt^2} \cdots a_n \frac{d^n x}{dt^n}$$

$$= b_0 y + b_1 \frac{dy}{dt} + b_2 \frac{d^2 y}{dt^2} \cdots b_n \frac{d^n y}{dt^n}. \quad (3.1)$$

The important points about this equation are: (1) it is "linear", because none of its terms involve powers or products of the dependent variable y. (2) It is "time invariant", because the coefficients $a_o \ldots a_n$ and $b_o \ldots b_n$ are constants. (3) It is an nth "order" equation, because n is the order of the highest differential coefficient occurring in it. In general, any differential equation can be solved by a number of integrations and the number of integrations necessary is the same as the order of the differential equation. Because each integration introduces a constant of integration the solution of an equation of order n will contain n such constants. For example

$$\frac{d^2 x}{dt^2} = a \tag{3.2}$$

first integration $\qquad \dfrac{dx}{dt} = at + c_1 \tag{3.3}$

second integration $\quad x = \frac{1}{2}at^2 + c_1 t + c_2 \tag{3.4}$

where c_1 and c_2 are the constants of the first and second integrations. Much of classical control theory involves manipulation of differential equations and for futher information the reader is referred to Brown (1965) and Milhorn (1966). For the present purposes only elementary knowledge of differential equations is necessary.

As explained in Chapter 1, equations containing time functions are generally transformed into operational notation and in block diagrams this generally takes the integral form. Thus transforming eqn. (3.2)

$$a = \frac{d^2 x}{dt^2} \quad \text{gives} \quad a(s) = s^2 x \tag{3.5}$$

and in the integral form

$$x(s) = a/s^2. \tag{3.6}$$

This equation is represented in block diagram form in Fig. 3.4. The input, $a(s)$, passes through two integrators to give the output $x(s)$. In order to specify the value of x in real terms, it is necessary to take account of the "initial condition" of each of the integrators. The initial conditions correspond to the states represented by the two

Fig. 3.4. Block diagram representation of eqn. 3.6.

constants of integration in eqn. (3.4). In practice, the initial conditions are easily incorporated into the flow graph analysis of a system and this procedure is explained later in this chapter.

In the application of differential equations to control systems, it is useful to make a distinction between state variables and rate variables. A "state variable" is one of a set of n variables, knowledge of which is sufficient to describe completely the behaviour of the system. Thus, an nth order system is described by a collection of n state variables and the state of the system is defined by the minimum number of such variables which, together with the system inputs, is sufficient to determine the behaviour of the system for all future time (Schultz and Melsa, 1967; Elgerd, 1967). Rate of change of state is measured in terms of rate variables, so that rate variable $= f$ (state variable).

In the block diagram representation of a system, the state variables correspond to the outputs of integrators. The number of state variables necessary to completely describe the system is generally referred to as the "order" of the system. Thus a first-order system has one essential state variable, a second-order system, two, etc. Essential state variables are those which are incorporated in the most parsimonious description of the system, such as that provided by the overall transfer function. (See however section 6.1.2.) A system may in practice be portrayed in a manner that reflects its physical composition, or in a more economical manner, relating only to the behaviour of the system. For example, the water content of the blood, and of the gut, are both the result of storage processes and both can be represented by state variables within the thirst system, as illustrated in Fig. 3.5a. These two storage processes act in parallel, as portrayed in this figure, and can thus be combined and represented by a single integrator, as in Fig. 3.5b. This rearrangement makes no difference to the transfer function of the sub-system, but the first arrangement is more meaningful physiologically. Nevertheless, both figures represent a first-order sub-system and the separation of the two storage components is merely a matter of convenience.

It is important to realize that the time-unit employed in analysis will affect the order of the abstract system. For example, a time-unit of one day, employed in investigation of the control of drinking, avoids not only complications due to circadian rhythms but also serves as a general strategy for simplifying the problem. In using a large time-unit the order of the system is reduced, because quick-acting processes can be ignored. The delay due to absorption and the transients involved in ingestion cannot be detected in an

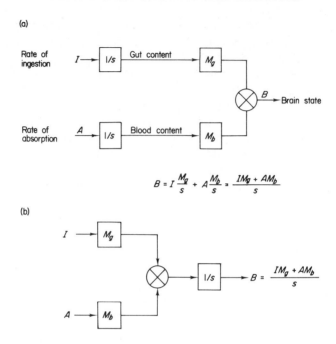

Fig. 3.5. Storage processes acting in parallel (a) are combined into a single process (b) mathematically equivalent, though less meaningful physiologically.

analysis based upon daily measurements. Consequently, the mechanisms responsible for these processes can be regarded as zero-order sub-systems and represented by constant parameters. As a first step in the analysis of complex systems this strategy has the advantage that the relationships between the slow-acting, underlying, processes are analysed first, thus giving an overall picture of the system. Subsequent experiments, employing a smaller time-unit, can then be made to show up particular details and to test hypotheses based upon the overall picture. This method has been used with some success in the analysis of thirst as a motivational system (McFarland, 1972), and this example is discussed in some detail later in this chapter. Before going into such detail however, it is necessary to consider how the order of a system can be identified in practice.

3.2.2 Transient Input Functions

The responses of simple linear systems to standard transient inputs are well known, and such knowledge can be extremely useful in the formulation of a working hypothesis about the nature of the system. Thus, the response of a system to a standard input can often be recognized as being characteristic of a zero-, first-, or second-order system. Three standard transient inputs are commonly used.

(a) "The unit impulse", often designated $\delta(t)$, is the limit of a rectangular pulse of magnitude $1/T$ and duration T, as T tends to zero (see Fig. 3.6). The unit impulse is also called the Dirac function. The

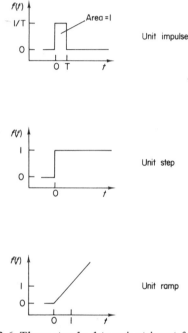

Fig. 3.6. Three standard transient input functions.

Laplace transform of the unit impulse is unity, viz. $\mathcal{L}\delta(t) = 1$. In practice, a short pulse can be considered as an ideal impulse when the system does not respond significantly during the pulse. For example, a quick intravenous injection of saline can be said to be an impulse input into the osmotic control system of the body fluids.

(b) "The unit step", $u(t)$, has amplitude $x = 0$ at time $t < 0$, and $x = 1$ at time $t > 0$ (see Fig. 3.6). The Laplace transform of the unit step, $\mathcal{L}u(t) = 1/s$, indicates that the unit step is the time integral of the unit impulse. In practice, the step function, which involves an infinite rate of change, can only be approximately realized. However, if the time-unit of measurement is of such a size that no response can be measured during the time interval in which the step occurs, then the input can be regarded as a step function for all practical purposes. A stimulus suddenly presented to an animal can often be regarded as a step input.

(c) "The unit ramp", $tu(t)$, is the time integral of the unit step (see Fig. 3.6). The Laplace transform of the unit ramp is thus, $\mathcal{L}tu(t) = 1/s^2$. In practice the ramp function is relatively easy to produce as it involves a variable which simply increases linearly with time.

3.2.3 Transient Responses

In his analysis of prey capture in mantids (section 2.3), Mittelstaedt (1957) explicitly states that he is interested in "the final steady-state of the system only—that is the position after all movements have come to rest, all actions and forces being in complete equilibrium". By deliberately ignoring the dynamic behaviour of the system, Mittelstaedt is able to construct a zero-order abstract system which is perfectly adequate for the purpose of erecting a working hypothesis. Strictly, a zero-order system cannot exist because every system must take some time to adjust to a change in input. However, where the time involved is very short, as is the case with the mantid's strike, the system can often be considered to approximate to a zero-order system. Obviously, such considerations will depend partly upon the time-unit used in measurement. The response of a zero-order system is a direct reflection of the input function, only scaling factors being altered. Thus the response to a step is itself a step, as illustrated in Fig. 3.7.

The response of a first-order system to a step input is exponential in form. McFarland and Budgell (1970a) subjected doves to a sudden

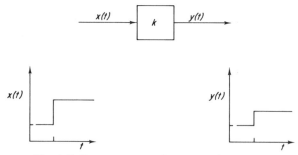

Fig. 3.7. Step-response of a zero-order system.

change in ambient temperature and measured the feather response by the method outlined in section 3.1.2. The ambient temperature was changed from $20°C$ to $40°C$ at a rate of $40°C$ per minute, and held at $40°C$ for the rest of the observation period. This sudden change in input approximates to a step function and the response of the system is illustrated in Fig. 3.8.

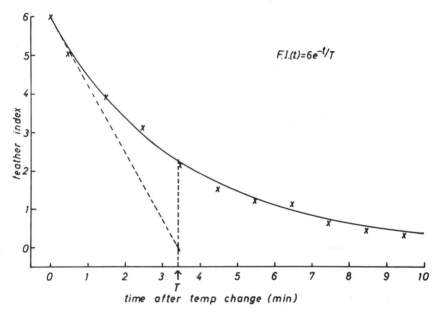

Fig. 3.8. Step-response of a first-order system. Observed feather index following a $20°C$ step increase in ambient temperature. T = time-constant of the feather response. (From McFarland and Budgell, 1970a.)

The function illustrated in this figure is an exponential function, of the type Ae^{-at}. A is the asymptote, or initial value, of the function, and $1/a$ is the time-constant of the function. The "time-constant" corresponds to the point on the time axis at which a straight line corresponding to the initial rate of increase cuts the asymptotic value of the independent variable. This is also the time when the independent variable reaches 63.4% of its final value.

On the basis of the discussion in section 3.2.1, we would expect that a system showing a characteristic first-order response can be described in terms of a block diagram having a single integrator. This, and other considerations, led McFarland and Budgell (1970a) to propose the block diagram, illustrated in Fig. 3.9, as a working model of the system controlling feather position in doves. In this block diagram there is a single integrator relating to the thermal capacity of the animal. Having set up a working hypothesis such as this, the first step in its verification must be to show mathematically how the system relates to the observed behaviour. Here the overall transfer function of the system is as follows.

$$\text{Transfer function} \quad F/H_{p-e}\,(s) = \frac{\phi_b\,\phi_h}{Cs + K_f T_d \phi_b\,\phi_h} \tag{3.7}$$

$$F(s) = H_{p-e}\,\frac{\phi_b\,\phi_h}{Cs + K_f T_d \phi_b\,\phi_h} \tag{3.8}$$

$$F(t) = \frac{H_{p-e}}{K_f T_d}\,(1 - e^{-K_f T_d \phi_b \phi_h / C.t}) \tag{3.9}$$

Thus at steady-state

$$F(t) = \frac{H_{p-e}}{K_f T_d}.$$

In other words, feather index is a linear function of ambient temperature. In a previous study (McFarland and Baher, 1968) this has been found to be true for the temperature range $6°-35°C$ (see Fig. 3.2). A step decrease in T_d will produce an increase in the final steady-state value of F (a decrease in the observed feather index, as the feathers become more sleeked), and from eqn. (3.9) it can be

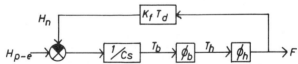

Fig. 3.9. Block diagram representation of the system underlying the feather response. H_{p-e} = heat production rate minus rate of evaporative heat loss. H_n = rate of non-evaporative heat loss. T_b = mean body temperature. T_h = hypothalamic temperature. F = feather index. T_d = temperature difference between T_b and the ambient temperature. K_f is a constant relating to the insulating properties of the feathers. C = thermal capacitance. ϕ_b and ϕ_h are unknown transfer functions. N.B. $F = 12 - FI$, the inverse of the empirical feather index. (From McFarland and Budgell, 1970a.)

seen that this value will be approached exponentially with a time constant = $C/K_f T_d \phi_h \phi_b$.

The second stage in the verification of a model of this type is to conduct further experiments, and check the results against those that would be expected theoretically. In the present case it was found that the time-constant of the feather response to a step-change in temperature was altered by food and water deprivation (Fig. 3.10). On the basis of the model, the time-constant of the change in hypothalamic temperature would be expected to alter accordingly, and this was found to be so (Fig. 3.11). In fact, it was found that the relationship between hypothalamic temperature and feather index was a linear one, and that the value of the relevant transfer function was changed by food and water deprivation (Fig. 3.12). This finding suggests that the changes in thermoregulatory function, induced by deprivation, are due to active CNS control, rather than the result of systemic physiological changes.

The work of Stark and his colleagues, on the human stretch reflex (Stark, 1968), provides an example of a transient response that is characteristic of a second-order system. The stretch reflex of the *pronator teres* muscle, which acts to resist disturbances causing supination, was studied by means of the apparatus illustrated in

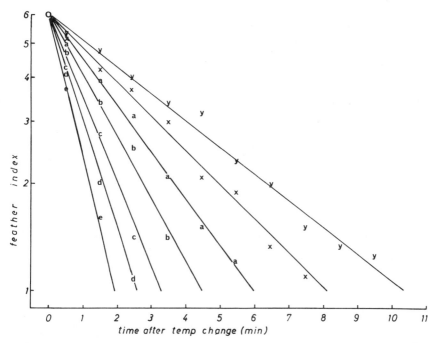

Fig. 3.10. Observed feather index as a function of time following a 20°C step-increase in ambient temperature. Data for doves tested after 0 h (a), 24 h (b), 48 h (c), 72 h (d), and 92 h (e) water deprivation; and 24 h (x), and 48 h (y) food deprivation. (From McFarland and Budgell, 1970a.)

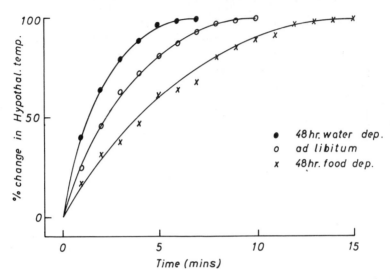

Fig. 3.11. Change in hypothalamic temperature as a function of time following a step-change in ambient temperature. (From McFarland and Budgell, 1970a.)

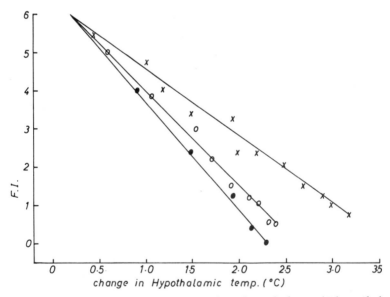

Fig. 3.12. Observed feather index as a function of change in hypothalamic temperature. Data from doves tested under *ad libitum* conditions (open circles), 48 h water deprivation (closed circles), and 48 h food deprivation (crosses). (From McFarland and Budgell, 1970a.)

Fig. 3.13. The subject is required to keep the pointer on target, while an impulse torque is delivered by means of a weighted pendulum which the subject is not able to see. The subject's wrist position is monitored by means of a potentiometer fixed to the shaft. In this experiment the shaft becomes part of the postural control system, contributing to its inertia, and this is taken into account in the analysis of the results.

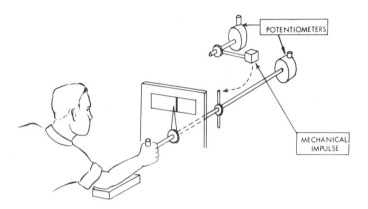

Fig. 3.13. Apparatus for measuring pronation and supination responses to mechanical disturbance. (After Milhorn, 1966.)

The results of this type of experiment are markedly affected by the tension with which the subject grasps the handle. This tension is under voluntary control and the *extensor digitorum communis* is the main muscle involved. The tension applied during each trial is monitored, either electromyographically, or by means of a sphygnomanometer cuff attached to the forearm. The change in wrist position following an impulse of torque is illustrated in Fig. 3.14, for low and high degrees of tension. The response consists in a gradually decaying oscillation, illustrating a phenomenon called "damping". The effect of increased tension is to decrease the amplitude, and increase the frequency, of the oscillation.

A response of this type is characteristic of a second-order system in which the state of the two integrators varies in an alternating fashion. The analysis of second-order transients is considerably more involved than that of first-order systems and the reader is referred to the next chapter for a more detailed discussion of the principles

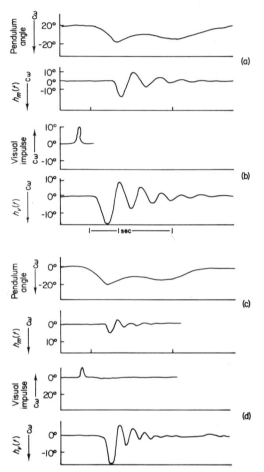

Fig. 3.14. Impulse response of a human stretch reflex. (After Stark, 1968.)

involved. For the present, it is sufficient to note that the basic characteristics of a system can often be recognized by casual inspection of the response to a standard transient input. More detailed analysis must await further investigation.

Although the techniques of transient analysis are relatively straightforward, their application does require a certain amount of knowledge and experience. A number of texts (e.g. Milsum, 1966; Milhorn, 1966) contain useful accounts of the procedures used in transient analysis and it is not necessary to cover this material here. However, it may be helpful to consider a particular example which serves to illustrate a number of the points raised in this chapter.

3.3 CONTROL OF DRINKING AS A FEEDBACK PROBLEM

The system maintaining water balance in birds and mammals can be represented, at the gross level, by a first-order system having two, essentially parallel, feedback loops (Fig. 3.15). McFarland (1965c)

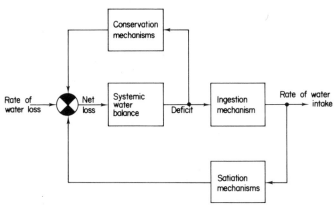

Fig. 3.15. Outline of system maintaining water balance.

showed that a system of this simplicity could adequately account for day to day changes in body weight and water intake during, and following, water deprivation in the Barbary dove *Streptopelia risoria*. To account for the effect of thirst on feeding, and for changes in drinking behaviour on a smaller time scale, it has proved necessary to "dissect" this simple system into its various sub-systems. Nevertheless, the strategy of employing a large time-unit in the initial analysis, and so obtaining a gross outline of the system, has proved to be a useful and meaningful one.

The negative feedback processes involved in the regulation of drinking can conveniently be divided into the physiological, which are mainly concerned with water conservation; and the behavioural, which are involved in obtaining water. Conservation mechanisms are of prime importance in the deprived animal. A rat or dove deprived of water is able to reduce its rate of water loss by means of a number of water conservation mechanisms. The pituitary-kidney antidiuretic mechanism is the best documented of these, and it is known that the antidiuretic hormone secreted by the hypothalamus-pituitary complex is capable of affecting a decrease in amount, and an increase in concentration, of excreted urine. In addition antidiuretic hormone probably acts to decrease the volume, and increase the concentration, of saliva; and to promote the absorption of water from the gut into the blood. These actions of antidiuretic hormone are thought to be more effective in mammals than they are in birds (Bartholomew and Cade, 1963), but some birds are able to reduce respiratory water loss during water deprivation (Cade, 1964). The mechanisms involved in this pulmocutaneous water regulation are not understood, though recent work indicates that the ability can be abolished by hypothalamic lesions (Wright and McFarland, 1969). In addition to direct water conservation, water can be saved indirectly by reduction of food intake during water deprivation. This can have the effect of mitigating obligatory excretory and thermoregulatory water loss (McFarland and Wright, 1969). The various mechanisms of water conservation act in parallel and their combined negative feedback effect results in deceleration of the rise in thirst during water deprivation. McFarland (1965c) found this function to be an exponential and by quantitative determination of food and water recovery curves, after specified deprivation schedules, was able to assess the relative importance of the different conservation mechanisms. In general, the degree of thirst is directly related to the effectiveness of the negative feedback processes involved in conservation.

It is a *sine qua non* of homeostasis that the CNS be able to monitor the state of the blood though this was not appreciated by early investigators. Hypothalamic sensitivity to changes in temperature (Hammel, 1965), osmosity (Cross and Green, 1959) and blood glucose (Anand *et al.,* 1962) has been demonstrated, and the roles of these variables in the regulation of feeding and drinking implicated (Brobeck, 1960; Andersson, 1953; Mayer, 1952). The means by which the brain obtains information concerning water and energy balance are not fully understood, though it appears that a number of different mechanisms act in parallel.

In addition to regulation of the internal environment, the CNS acts as a controller in negative feedback loops which encompass the external environment. In particular, hunger and thirst differ from some other aspects of homeostasis in that appropriate behaviour is essential for maintenance of the status quo. The deprived animal cannot compensate for loss of water or energy, it can only reduce the rate of loss. Thus feeding and drinking behaviour can be regarded as an essential part of homeostasis and can be examined from the point of view of feedback theory.

Drinking behaviour is thought to be initiated by changes in

systemic osmotic pressure and other factors. Water intake is probably monitored at three levels–oral, alimentary and systemic–and there appear to be considerable species differences in the weighting that is given to the various factors (Adolph, 1950). Satiation is thus a multifactor process, but from the feedback theory viewpoint the various satiation mechanisms act in parallel and provide negative feedback by means of which the amount of water consumed at any particular time is regulated in accordance with a "command" set up by the monitored systemic deficit, as illustrated in Fig. 3.16. This

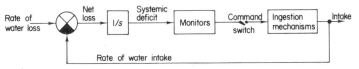

Fig. 3.16. Outline of thirst satiation system.

simplified interpretation of the situation will serve as a starting point for our analysis.

3.3.1 Satiation as a Feedback Process

The system outlined in Fig. 3.16 is portrayed in more compact form in Fig. 3.17. In this block diagram the input, the rate of loss L,

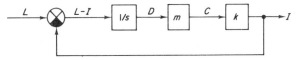

Fig. 3.17. Block diagram of thirst satiation system. L = rate of water loss. D = systemic deficit. C = thirst command. I = rate of water intake. m and k are system parameters.

is considered to be constant at all times, since it is assumed that the animal is maintained under constant environmental conditions. The value of this input variable, $L = \alpha(t)$ is written as α/s in Laplace notation, for it is assumed that a step-function occurred when the input was first applied. The difference between the rate of loss L and the rate of intake I is the net loss $L - I$, which is integrated to give the systemic deficit, D. The deficit is monitored by sense organs that are assumed to have constant properties and are thus represented by the constant parameter m. The monitored deficit is represented neurally in the CNS, and gives rise to a command, C, which actuates the ingestion mechanism. Although the ingestion is itself a complex system, it is represented by the single parameter k. This over-simplification is justified on the grounds that the transients involved in this part of the system are very fast compared with the time scale of the present analysis. It is thus permissible to represent the ingestion mechanism as a zero-order sub-system. The output of the ingestion mechanism is the rate of water intake I, which feeds back to reduce the deficit. This representation of the mechanism controlling water intake can thus be seen to be a simple, first-order feedback system.

When the animal is deprived of water the ingestion mechanism can no longer operate and this is equivalent to opening a switch at the command level. The ingestion mechanism thus suffers a step-change in input. When water is provided, after deprivation, the command switch is closed and there is a step-change in the opposite direction. In considering the behaviour of the system during deprivation and recovery, it is necessary to take into account the state of the integrator at the time at which each step-change is made. This is equivalent to specifying the initial conditions in a differential equation. In terms of operational calculus, it is convenient to account for the initial conditions by employing the flow graph method.

The first step is to convert the block diagram (Fig. 3.17) into its flow graph equivalent (Fig. 3.18). The value of the single state variable in the system is that of the output of the integrator, i.e. the deficit D. This can be determined directly by flow graph simplification, as shown in Fig. 3.18. If the system were of higher order, it would be necessary to make a similar determination of the value of each of the state variables. The number of these will be the same as the order of the system (see section 3.2.1).

From Fig. 3.18 it can be seen that the steady-state value of

$$D_{(s)} = \frac{\alpha}{s(s + km)} \quad \text{and} \quad D(t) = \alpha/km$$

(see Table 1.2). When the animal is water deprived, the feedback loop is broken and the value of D increases as the water loss is integrated. According to the present model, the value of D will rise linearly under these conditions, but in practice this rate of increase decelerates due to the action of the conservation feedback loops, which have been omitted from the present considerations in order to simplify matters. The value of D will continue to rise until water is again made available to the animal (Fig. 3.19). At this point, the value of D, called D', is the initial condition which must be taken

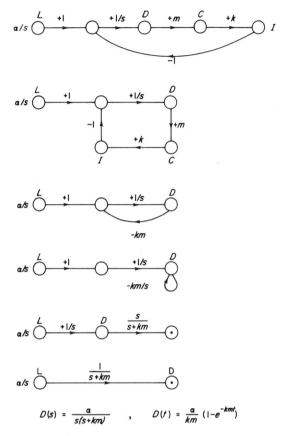

$$D(s) = \frac{a}{s(s+km)} \quad , \quad D(t) = \frac{a}{km}(1-e^{-kmt})$$

Fig. 3.18. Flow graph equivalent of block diagram in Fig. 3.17. Stages in flow graph reduction are illustrated.

into account in calculating the course of recovery from deprivation.

The first step in the procedure for calculating the recovery function, is to construct a flow graph representing the configuration of the system during the recovery period, i.e. with the switch closed. The initial condition of D is incorporated into this flow graph as an extra input, as shown in Fig. 3.20. The value of I is then calculated from the flow graph by the normal method (Fig. 3.20), and from this it can be seen that

$$I(s) = D' \frac{mk}{s+mk} + \frac{\alpha mk}{s(s+mk)} \tag{3.10}$$

$$\mathcal{L}^{-1} I(s) = I(t) = D' mke^{-mkt} + \frac{\alpha mk}{mk}(1-e^{-mkt}) \tag{3.11}$$

Equation (3.11) is composed of two exponential functions added together. This sum is illustrated graphically in Fig. 3.19, in which the two dotted curves are added to give the resultant water intake recovery curve, which is exponential in form, as might be expected from a first-order system.

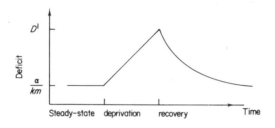

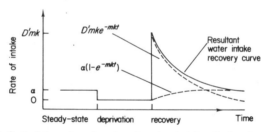

Fig. 3.19. Deficit (above) and rate of water intake (below) as a function of water deprivation and recovery. Dotted curves indicate components of the recovery curve, as calculated in Fig. 3.20. D' = value of deficit at the end of deprivation.

Before going on to consider further behavioural aspects of satiation it might be useful to discuss some general points arising from this simplified analysis of satiation as a feedback process. Firstly, it should be noted that in flow graph analysis of first-order systems the procedure is identical to that for zero-order systems, illustrated in sections 1.4 and 1.5. In particular, the Laplace variable s is treated as an algebraic constant and the necessary conditions for flow graph simplification are not violated by inclusion of time-functions. The only additional procedures that are necessary in the analysis of first, and higher, order systems are the transformation of resultant equations from Laplace-functions to time-functions and the incorporation of initial conditions. The procedure for incorporating initial conditions by flow graph methods is outlined in greater detail by Naslin (1965), McFarland (1965d) and by Huggins and Entwisle (1968). More general treatment of initial conditions is contained in Milsum (1966).

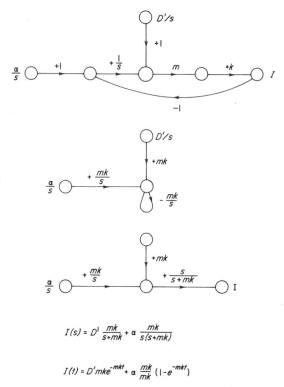

$$I(s) = D' \frac{mk}{s+mk} + a \frac{mk}{s(s+mk)}$$

$$I(t) = D'mke^{-mkt} + a \frac{mk}{mk}(1-e^{-mkt})$$

Fig. 3.20. Flow graph illustrating a method of incorporating initial conditions, prior to flow graph reduction.

A second point is that, in this example, step-changes are induced by suddenly changing the value of a parameter rather than the value of an input variable. As the properties, or configuration, of the system are changed by this procedure it is necessary to divide the mathematical analysis into distinct parts relating to the time periods before, and after, the step-change. The initial conditions must be taken into account for each part separately.

Finally, the present model predicts that the course of recovery from water deprivation, i.e. the satiation curve, is exponential in form. The next step is to verify this prediction experimentally.

3.3.2 Short-term Control of Water Intake

Daily measurements of water intake in doves, during recovery from water deprivation, yield exponential recovery curves for periods of deprivation up to 96 h (McFarland, 1965c). Thus, in an analysis based upon a time-unit of 24 h, a first-order model is adequate. But in a more short-term analysis the delay involved in the absorption of water from the lumen of the gut into the blood must be taken into account. As the system stands (Fig. 3.17), ingested water affects the drinking rate solely through its ultimate effect on the body fluid. Any delay in this process would prolong the time spent drinking unless a separate mechanism for stopping drinking were present. In the long-term analysis this problem is circumvented by considering drinking as a continuous function of time. Physiologists have long recognized the importance of such a stop-mechanism, and Oatley (1967), after reviewing the relevant experiments, introduces such a mechanism into his control model for the physiological basis of thirst. Before discussing the details of the stop-mechanism, however, it will be instructive to consider the nature of the absorption delay.

Oatley (1967), noting that in the response to an injection of water into the rat's stomach the total water absorbed from the gut rises exponentially with a time-constant of about 20 min (Adolph and Northrop, 1952), suggests that the absorption delay is in the form of an exponential lag. In birds one might expect the situation to be complicated by the presence of the crop, but Dicker and Haslam (1966) found that the water content of the crop of domestic fowl decays exponentially and that there is a coincident peak of diuresis, indicating that water absorption must have kept pace with the emptying crop. Measurement of electrolyte concentration in doves, following ingestion of water, indicates that water is absorbed exponentially with a time-constant of about 5 min (McFarland and Perinchief, 1971). Thus it appears likely that the absorption delay is exponential in form in doves as well as in rats.

The required delay system follows naturally from the assumption that the rate of absorption A is proportional to the gut water content. Figure 3.21 shows that a first-order sub-system results when the abosrption rate is subtrated from the rate of inflow, giving a net rate of inflow which is integrated to give gut content. Thus in response to a step intake of water the rate of absorption $A(t) = x(1 - e^{-ht})$, where x is the height of the step and the time-constant is determined by the value of the absorption constant h. In reality the system is much more complex than this, as is shown by more detailed studies on the rat (Toates and Oatley, 1970).

An absorption delay of this simple type is incorporated by McFarland and McFarland (1968) in their model for short-term control of drinking in the Barbary dove. A version of this model is illustrated in Fig. 3.21. Ignoring for the moment the oral feedback loop, the system can be described by the transfer function

$$I/L(s) = \frac{k(s+h)}{s^2 + s(h+kg) + hfk} \tag{3.12}$$

which is characteristic of a second-order system. The physiological implications of this equation are to be seen in the nature of the

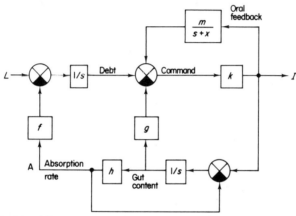

Fig. 3.21. Simplified version of McFarland and McFarland's (1968) system for the short-term control of the drinking response. L = rate of water loss, I = rate of water intake, k = parameter of the ingestion mechanism, g = that of gut-inhibition mechanism, h = absorption constant, and f the hydration coefficient.

parameters of the system. The absorption constant h determines the time-constant of the absorption delay. The parameter f represents the "hydrating power" of the absorbed water and can be taken to vary with the salinity of the water ingested. Thus saline would be a less effective hydrating agent than distilled water. The effectiveness of the gut-inhibition system is determined by the parameter g, which represents the combined effect of the sense organs involved in the measurement of gut content and of the calibration of the messages in the brain. Finally, the ingestion mechanism is represented by the parameter k as in the previous model (section 3.3.1). McFarland and McFarland (1968) supposed that the value of g is calibrated to match the salinity of the water that is generally available. In particular, if

$f = g$, and this value is called j, then from eqn. 3.12

$$I/L(s) = \frac{k(s+h)}{(s+h)(s+kj)} = \frac{k}{s+kj}.$$ (3.13)

In response to a step-function in D, the water debt, such as would occur when water is presented after a period of deprivation the rate of water intake would decline exponentially to its normal level, and the amount of water drunk during the first drinking bout of recovery from deprivation is given by the following equation

$$I/s = \frac{D}{s} \cdot \frac{k}{s+kj}, \qquad \mathcal{L}^{-1}I/s = \int Idt = \frac{D}{j}(1-e^{-kjt}).$$ (3.14)

Here we have an example of a second-order system behaving like a first-order system due to the action of internal feedback. As will be explained in the next chapter, the response of a second-order system depends upon the "damping factor". A system which is "critically damped" has a response that is similar to an exponential.

An animal which is accustomed to drinking pure water will presumably rely on its gut-inhibition mechanism to determine the quantity of water necessary for restoration of water balance. If such an animal is given saline, which is below the taste rejection threshold, the quantity of water ingested will not be sufficient to restore the water balance. When the animal is drinking at a slow rate, as in an operant situation, ingestion of saline will have the effect of increasing the damping of the system because the value of the parameter f will be lower than normal and will no longer match the calibrated value of g. In addition to altering the system damping the amount of water ingested will also be altered. The value of the new asymptote of the satiation curve can be used to calculate the new value of f and this value can be substituted in the appropriate equation to generate a specific, experimentally verifiable, prediction. Thus eqn. 3.12 can be rewritten as follows

$$I/L(s) = \frac{k(s+h)}{s^2 + s(h+kg) + hfk} = \frac{k(s+h)}{(s+a)(s+b)}$$ (3.15)

where $a + b = h + kg$ and $ab = hfk$. In response to a step-function in D,

$$I/s = D\frac{k(s+h)}{s(s+a)(s+b)} \quad \text{and} \quad \int Idt = D\frac{kh}{ab} - \frac{h-a}{a(b-a)}e^{-at} - \frac{h-b}{b(a-b)}e^{-bt}$$

At steady-state

$$\int Idt = D\frac{kh}{ab}(t) = \frac{D}{f}(t).$$ (3.16)

As the normal asymptote of the drinking response is $D(t)$, i.e. the animal drinks to make up its water debt, the value of f can be obtained from the equation

$$f = A_w/A_s \tag{3.17}$$

where A_w is the amount of water consumed, and A_s is the amount of saline consumed, under identical conditions.

As an empirical test of these predictions, McFarland and McFarland (1968) trained five adult doves to obtain water by pecking at an illuminated key in a Skinner box. Rewards of 0.1 cm³ water were delivered to the subject for each peck. The purpose of this procedure was to slow down the drinking response and make it more amenable for study. During testing, each subject was deprived of water for 48 h before each session, and was then run until it reached a satiation criterion of 5 min without pecking at the key. Each subject was tested with water reward and with rewards of 0.5%

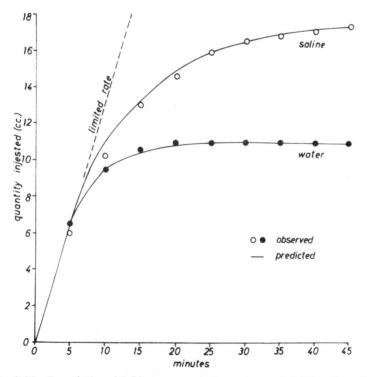

Fig. 3.22. Cumulative drinking response for water and 0.5% saline. (From McFarland and McFarland, 1968.)

saline which is below the rejection threshold for pigeons (Duncan, 1960). The results show that whereas the average consumption of water is 11.0 cm^3, the average consumption of saline is 17.5 cm^3. Substitution of these values in eqn. 3.17 makes it possible to predict quantitatively the satiation curves under the two conditions. Comparison of the observed and predicted satiation curves is illustrated in Fig. 3.22.

3.3.3 Separation of Oral and Gut Feedback Mechanisms

For the purpose of considering the uninterrupted drinking response the oral and gut mechanisms act in parallel and their combined effect can be represented by the single parameter g (Fig. 3.21). However, the roles of the two mechanisms can be distinguished by experimental interruption of the drinking response.

As mentioned above, the response of the gut-inhibition mechanism to a step intake of water is determined by the response of the absorption mechanism. The command is decremented by an amount proportional to the water content of the gut and also by the amount of water absorbed. As these two quantities are complementary, the command is a measure of the amount of water ingested, and is independent of the state of the absorption mechanism when the system is critically damped. But the command is only a true measure of the water needs of the animal in the absence of oral factors. If the drinking response were interrupted before satiation was reached, drinking would be resumed at the same rate as that at which it was terminated only if oral factors are in operation.

In Fig. 3.21 oral factors are incorporated into the system as a decaying feedback loop acting on the command. The time constant of the decay is determined by the parameter x, and it is reasonable to assume that this parameter has the same value as h, the absorption constant. If the values of these parameters were not the same the drinking response would show a discontinuity which is not apparent in the results illustrated in Fig. 3.22.

In calculating the transfer function of the system portrayed in Fig. 3.21, we have to consider the possibility that the oral factors act as a negative, or as a positive feedback mechanism.

$$I/L = \frac{k(s+h)(s+x)}{s(s+h)(s+x) + k(s+x)(fh+gs) \pm kms(s+h)} \qquad (3.18)$$

Let $x = h$,

$$I/L = \frac{k(s+h)}{s(s+h)+k(fh+gs)\pm kms} = \frac{k(s+h)}{s^2+s(h+k)(g\pm m)+kfh} \qquad (3.19)$$

Let $f = g \pm m = j$,

$$I/L = \frac{k(s+h)}{s^2+s(h+kj)+kjh} = \frac{k(s+h)}{(s+h)(s+kj)} = \frac{k}{s+kj} \qquad (3.20)$$

Clearly, the introduction of oral factors makes no difference to the uninterrupted drinking response. The effect of interrupting the drinking response will depend upon the nature of the oral feedback. Suppose that, 10 min after the start of an operant drinking response, the availability of water was terminated for a period of 15 min after which access to water was permitted again. During this 15 min period the decline of gut inhibition is compensated by the absorption of water, but the adaptation of oral factors is not so compensated. Consequently, negative feedback from oral factors will decline, causing a concomitant rise in the command. Conversely, a decline in positive feedback from oral factors would cause the command to decline during interruption of drinking. Thus, negative feedback from oral factors will result in drinking being resumed at a higher rate than that at which it was terminated, while positive feedback will result in a lower resumed rate of drinking.

As a simple test for the role of oral factors in the control of drinking, five birds were tested in the manner described in the previous experiment. Each subject was given an interrupted and a non-interrupted test, according to a Latin square design. Drinking was interrupted by turning off the response-key and water delivery mechanism for periods of 5 or 15 min. The response rate was recorded as a continuous function of time. The results, shown in Fig. 3.23, indicate clearly that drinking after interruption is resumed at a lower rate than that at which it was terminated. These results suggest that oral factors provide a positive feedback which presumably serves to maintain the drinking response until it is terminated by gut-inhibition.

Further evidence for the view that oral factors provide positive feedback comes from the studies of Wiepkema (1968) on the feeding behaviour of CBA mice. Wiepkema found that the length of bouts of uninterrupted feeding behaviour is an increasing function of length of previous food deprivation. Bout length is thus an indication of hunger level. However, the initial phase of a normal meal is characterized by a progressive increase in bout length, suggesting that hunger is increasing during this period and that positive feedback is

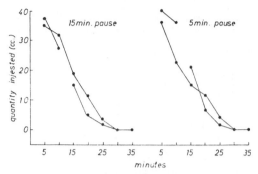

Fig. 3.23. Rate of drinking under normal conditions (continuous curve) and when interrupted for 5 or 15 min (broken curve). The break in the curve occurs at the point at which drinking was interrupted. (From McFarland and McFarland, 1968.)

operating. When the mice were given unpalatable, but nutritionally unaltered, food, the increase in bout length was much less marked. This result suggests that the degree of positive feedback was lessened by reducing the effectiveness of oral factors. From a study of feeding behaviour in rats, Le Magnen (1968) found a similar effect of palatability on ingestion rate, and he concluded that oropharyngeal stimulation at the beginning of a meal is involved in a positive feedback mechanism which is later counteracted by satiating effects of gastric origin.

In addition to satiation, the various monitoring mechanisms probably play some role in providing reinforcement cues. Miller and Kessen (1952) found that milk taken normally by the mouth is a much more powerful reinforcer than milk injected directly into the stomach. On the other hand, rats will learn to bar press for intragastric food (Epstein and Teitelbaum, 1962) or water (Epstein, 1960) reward, although prolonged training is usually necessary to establish the behaviour. The satiating and rewarding consequences of drinking can be separated in doves with chronic oesophageal fistulae (McFarland, 1969b). Doves were tested in an operant situation to determine the relative roles of oral and alimentary factors in the reinforcement and satiation of drinking behaviour. In reinforcement tests subjects transferred from a non-rewarded key to a novel rewarded key only when reward was presented orally. It appears that alimentary factors do not reinforce such behaviour. In satiation tests rewards were presented simultaneously via the fistula and orally. Satiation depended on the total amount ingested irrespective of the proportions delivered via the two routes. These results can be interpreted in terms of the feedback mechanisms involved in satiation and reinforcement. Thus alimentary loading factors provide the negative feedback necessary for satiation, and oral factors provide positive feedback, which serves to maintain drinking in the face of competing stimuli from other motivational systems. In other words, oral factors serve to reinforce the ongoing behaviour and maintain its momentum until satiation is complete.

3.3.4 Discussion

Section 3.3 is concerned with a particular example of the application of transient analysis to a behavioural system. Before going on, in the next chapter, to a more advanced type of systems analysis, it may be useful to discuss some general points arising from this account.

The study of the control of drinking as a feedback problem, as outlined above, provides an example of the general strategy of progressively contracting the time-unit used in measurement and analysis. In the earlier studies (McFarland, 1965c), a 24 h time-unit was employed, partly to avoid complications due to circadian rhythms, and partly as a means of arriving at a general outline of the system. From this work it is apparent that the system controlling water balance can be described as a first-order system at a gross level. Transients induced by step-changes in water availability have large time-constants, which can be shown to be due to a number of feedback loops, acting in parallel and serving the function of water conservation. From this point the analysis can take two different paths. On the one hand, further investigation of the conservation mechanisms elucidates the basic interactions between the system controlling water intake, and those controlling food intake (McFarland, 1964; 1965a; McFarland and Wright, 1969) and body temperature (McFarland, 1967; Wright and McFarland, 1969; McFarland and Budgell, 1970b; Budgell, 1970). On the other hand, analysis on a more short-term basis provides more detailed knowledge of the structure of the systems controlling fluid balance (Pace, 1961; Reeve and Kulhanek, 1967; Oatley, 1967; Koshikawa and Suzuki, 1968) and water ingestion (McFarland and McFarland, 1968; Toates and Oatley, 1970).

Such short-term analysis takes account of the transients involved in the distribution of water between the body compartments, hormonal control of water and salt excretion, absorption of water from the intestine, etc. When very small time-units are employed problems due to non-linearities in the system can arise. Non-linearities due to the thirst-threshold, and to limitations on the ingestion rate, are considered by McFarland and McFarland (1968). In general, problems involving non-linearities cannot be solved by analytical methods and one solution is to "linearize" the system by employing a larger time-unit (see section 3.1.3), making such assumptions and simplifications as are consonant with the empirical data. A more satisfactory approach is to simulate the system, non-linearities and all, on an analogue or digital computer. With respect to thirst, the most progress to date has been made by Toates and Oatley (1970), and this work is discussed in Chapter 6.

Most studies on the control of drinking draw on current physiological knowledge in the formulation of new models and working hypotheses. This approach is essentially synthetic, as sub-systems of known performance are integrated to form a working model of the whole system. It is common in the application of control theory to physiology (Milhorn, 1966). However, many systems of a more behavioural nature are not open to this method of approach, generally because their physiology is not well understood. Thus while some degree of synthesis is often possible, particularly with systems whose performance is closely related to systemic physiology, sophisticated analytical methods are likely to be particularly fruitful in the investigation of behavioural systems.

CHAPTER 4

Behaviour as a Function of Frequency

Consideration of the behaviour of a system as a function of the frequency of a periodic input is a well-established method of systems analysis. This method, generally called "frequency analysis", has considerable advantages over the type of transient analysis discussed in the previous chapter. The response to a periodic input is itself periodic and can be assessed by repetitive measurements of a very simple nature compared with the accurate measurements required by transient analysis. The measuring procedure tends to average out variability in the behaviour of the system and thus leads to a more accurate picture of its dynamic performance. It is also possible to use a periodic input function of small amplitude, whereas transient inputs generally have to be of fairly large magnitude to produce a response that is sufficiently distinct to make accurate measurement possible. For this reason, such "small-signal analysis" is considered to be a more "natural" method of exciting a system.

In addition to purely practical advantages frequency analysis has a firm mathematical foundation which provides a conceptual base for more advanced forms of systems analysis. In this chapter an attempt is made to introduce the reader to the basic ideas of frequency analysis, without involving a formal mathematical exposition of the principles involved. In particular, methods such as the Nyquist plot, which require knowledge of the use of complex numbers, are not covered here. Readers requiring an introduction to the application of these methods to biological systems analysis should consult Milsum (1966) or Milhorn (1966).

4.1 FREQUENCY ANALYSIS

4.1.1 The Sinusoidal Input Function

The periodic function most commonly used in frequency analysis is the sine function. By trigonometric definition the value of $\sin \alpha$ is the ratio of the side opposite the angle α and the hypotenuse in a right-angled triangle (Fig. 4.1). If we imagine the line XY to be the radius of a circle, an arm rotating at a constant rate counter-clockwise around Y, the point X can be seen to trace out a

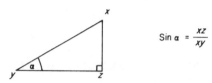

Fig. 4.1. Trigonometric definition of the sine of an angle.

sinusoidal function of time (Fig. 4.2). As α increases from zero, sin α rises to a maximum of unity at $\alpha = 90°$, reduces to zero at $180°$, drops to minus one at $270°$, and returns to zero at $360°$. The value of α at any time t can be defined as $\alpha = \omega t$, where ω is a constant, called the angular velocity or "angular frequency", measured in radians per second. Thus

$$x(t) = A \sin \omega t = A \sin 2\pi ft \qquad (4.1)$$

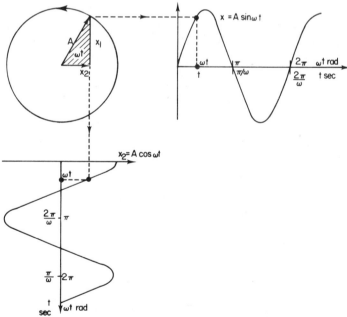

Fig. 4.2. Generation of sinusoidal waves from a rotating vector. (From Milsum, 1966.)

where f is the "frequency" in cycles per second and A is the "amplitude" of the sine wave (Fig. 4.3).

Two sine waves may thus differ in frequency and in amplitude, and they may also differ in "phase", expressed as the angular difference ϕ between waves of the same frequency (Fig. 4.3). Since it

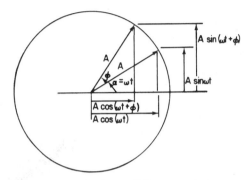

Fig. 4.3. Amplitude A, and phase difference ϕ, between two sinusoidal functions. (From Milsum, 1966.)

is not always true that $x = 0$ when $t = 0$, it is usual to write eqn. (4.1) in the more general form

$$x(t) = A \sin(\omega t + \phi). \tag{4.2}$$

The basic principle of frequency analysis is that a sinusoidal input applied to a linear system will produce a sinusoidal output of the same frequency, but differing in amplitude and phase in accordance with the nature of the system. The analysis is conducted on the basis of comparison of the amplitude and phase differences between input and output. These differences will depend, primarily, upon the number of integrators and differentiators within the system. The effect of these operations upon the sine wave is therefore of prime importance.

If $x(t) = A \sin \omega t$,

then

$$\frac{dx}{dt} = A\omega \cos \omega t = A\omega \sin(\omega t + \pi/2). \tag{4.3}$$

Thus differentiation of a sine wave produces a sine wave of the same frequency but differing in amplitude and exhibiting a positive phase shift of $90°$. A positive phase shift is called a "phase advance" or "phase lead". Further differentiation produces further similar changes in amplitude and phase, viz.

$$\frac{d^2 x}{dt^2} = A\omega^2 \sin(\omega t + \pi). \tag{4.4}$$

Integration of a sine wave has the opposite effect. Thus a single integration produces a 90° phase lag

$$\int x dt = \frac{A}{\omega} \sin(\omega t - \pi/2) \tag{4.5}$$

and a double integration produces a 180° lag

$$\int \int (x dt) dt = \frac{A}{\omega 2} \sin(\omega t - \pi). \tag{4.6}$$

4.1.2 Frequency Analysis in Practice

As an experimental procedure, frequency analysis consists in applying a sinusoidal signal to a suitable input and measuring the response at a suitable output. The input frequency is varied over a certain range and the relationship between input and output, with respect to amplitude and phase, is calculated as some function of the input frequency. The most commonly used amplitude relationship is the "amplitude ratio" AR, defined as the ratio of output and input amplitudes (Fig. 4.4). The phase relationship is generally the angular difference ϕ (Fig. 4.5).

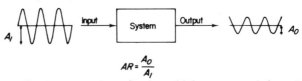

$$AR = \frac{A_O}{A_I}$$

Fig. 4.4. Amplitude attenuation of a sinusoidal wave passed through a system.

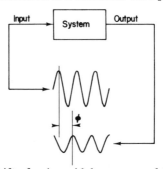

Fig. 4.5. Phase shift of a sinusoidal wave passed through a system.

The first problem in empirical frequency analysis is to obtain a suitable input function. Although sinusoidal functions are preferable analytically, they are not always easy functions to apply to biological systems. In some cases it may prove necessary to resort to other types of periodic input, such as repetitive pulses or triangular waves. In principle, the procedure involved in the use of such functions is similar to that for sinusoidal inputs but it is generally much more complicated in practice, because the sinusoid is the only waveform that reproduces itself when passed through a linear system.

If a sinusoidal input is applied to a system at rest, the output amplitude will not be constant until the effects of the input have been distributed evenly throughout the system. After the initial "warm-up" period, a sinusoidal steady-state is reached in which the output amplitude is constant from one cycle to the next. Thus the second step in empirical frequency analysis is to ensure that the initial transients have decayed before any measurements of amplitude or phase are taken.

The third step is to ensure that the system behaves in a linear fashion, at least under the prevailing experimental conditions. The response of a linear system to a sinusoidal input should itself be sinusoidal and a marked deviation from this form indicates that a non-linear mechanism is involved. Sometimes it is possible to avoid non-linearities by restricting the input to a particular mean value, producing a range of excitation within which the system behaves in a linear fashion. For example Budgell (1971), investigating behavioural thermoregulation, trained Barbary doves to peck at a key in a Skinner box to turn on a heater in a cooled environment. The trained animals were subjected to sinusoidal temperature changes while the heater was disconnected from the peck mechanism. Sample results from these experiments are illustrated in Figs 4.6 and 4.7. When the temperature varied sinusoidally about a mean temperature of 32.5° C, the response rate shows marked discontinuities which indicate that a threshold (non-linear) phenomenon is involved. When the mean temperature was 27° C, the discontinuities are no longer apparent and the threshold value is never reached.

Further points may be illustrated by reference to another example. In investigating the effects of ambient temperature upon drinking, McFarland and Budgell (1970b) made the following experiment:

Three Barbary doves *Streptopelia risoria* were trained to obtain 0.1 cm^3 water rewards by pecking at an illuminated key in a Skinner box which was installed inside a climatic cabinet. Rewards were

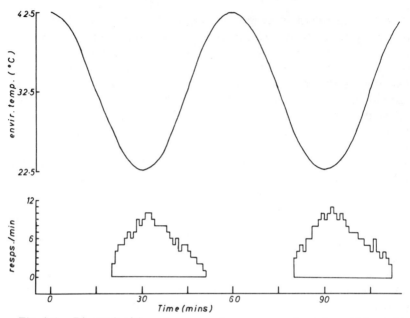

Fig. 4.6. Discontinuities in operant response to sinusoidal ambient temperature fluctuations indicate the presence of a threshold in the system. (From Budgell, 1971.)

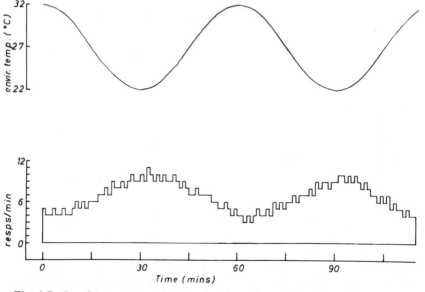

Fig. 4.7. Continuous operant response when sinusoidal ambient temperature fluctuations are below threshold. (From Budgell, 1971.)

delivered on a random interval schedule, having a mean interval of 2 min (VI 2 min). The animals were trained for ten weeks, until they responded at a constant rate for a 3 h period, following 48 h water deprivation.

During testing, the animals were run every two days and were deprived of water during the intervening periods. Tests were conducted alternately, at a constant temperature of 15°C and at a temperature fluctuating sinusiodally between 5°C and 25°C. Tests were run at frequencies of 0.40, 0.50, 0.66, 1.00, 1.33, 2.00 and 4.00 cycles/h. The accuracy of the sinusoidal temperature function was ±0.25°C. Pecks delivered and rewards obtained were recorded on an event recorder throughout the whole of each 3 h test period, but the records for the first and last 30 min were discarded to avoid possible effects of initial transients, and satiation, respectively.

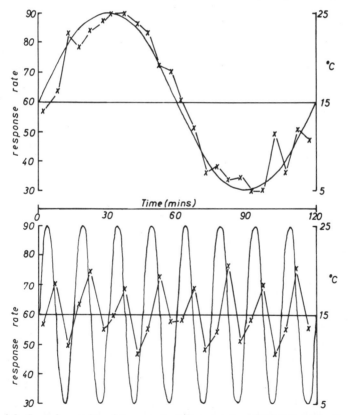

Fig. 4.8. Sample results of temperature frequencies of (a) 0.5 cycles/h and (b) 4 cycles/h. Response rate in pecks/5 min is indicated by crosses, ambient temperature by the continuous curve. (From McFarland and Budgell, 1970b.)

The results obtained from two sample test sessions, at different frequencies, are illustrated in Fig. 4.8. When the ambient temperature changes slowly the animal's response rate follows the input function with reasonable accuracy. But if the temperature changes at a high frequency the animal is unable to "keep up" and there is a noticeable phase lag and attenuation of response amplitude. At very high frequencies the output amplitude would approach zero and the animal would respond at a constant rate proportional to the average temperature. For technical reasons it was not possible to reach such frequencies with the apparatus available for this experiment.

The essence of frequency analysis is to determine the form of the function describing the manner in which amplitude ratio and phase angle progressively change as inputs of higher and higher frequency

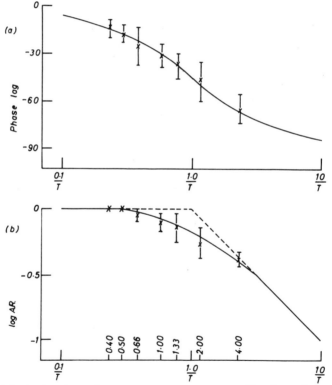

Fig. 4.9. Bode plot of mean response rate as a function of temperature frequency: (a) observed phase lag matched to that of a first-order system (continuous curve); (b) observed compared with expected (continuous curve) amplitude ratio. Vertical lines indicate the range. (From McFarland and Budgell, 1970b.)

$$e_c/e_{in} = \frac{1}{LCs^2 + RCs + 1} \tag{4.11}$$

In this case

$$\omega_n = \frac{1}{\sqrt{LC,}} \quad \text{and} \quad \zeta = \frac{RC}{2\sqrt{LC}}$$

In the absence of damping the step response of a second-order system is a sine wave, with average value determined by the height of the step and frequency called the "natural frequency" of the system. In the presence of damping these oscillations decay exponentially with a time-constant related to the damping ratio. Figure 4.10 illustrates the step response of a generalized second-order system, with various values of the damping ratio ζ. When $\zeta < 1$, the system is said to be "underdamped" and will overshoot the final steady-state

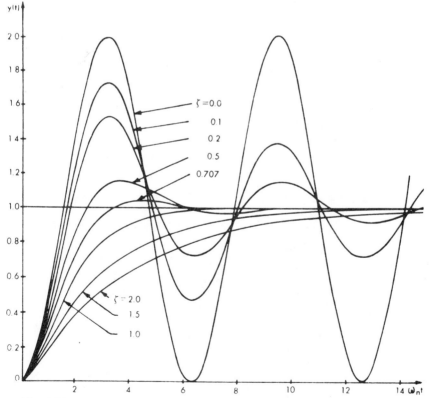

Fig. 4.10. Step response of a generalized second-order system at various values of the damping ratio ζ. (From Milsum, 1966.)

FEEDBACK MECHANISMS IN ANIMAL BEHAVIOUR

level. The lowest value of the damping ratio for which the overshoot just disappears is $\zeta = 1$, and such a system is said to be "critically" damped. When $\zeta > 1$, the system is "overdamped", and the more ζ approaches infinity, the longer the response takes to reach its final value.

In practice, second-order systems can be identified by their

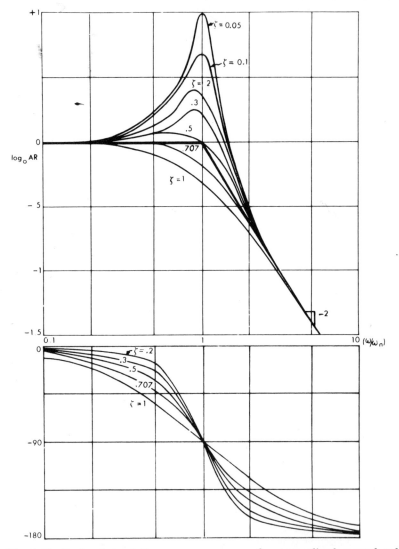

Fig. 4.11. Bode plot of frequency response of a generalized second-order system, at various values of the damping ratio ζ. (From Milsum, 1966.)

transient response, particularly when the damping is low, as shown by the impulse responses of the human stretch reflex illustrated in Fig. 3.14. When overdamped, however, the response of a second-order system becomes very similar to that of a first-order system and, unless the measurements are very precise, transient analysis becomes unreliable. Frequency analysis is a much more accurate method of identifying systems of this type.

The frequency response of a generalized second-order system is indicated by the Bode plot of Fig. 4.11. Note that as the input frequency approaches the natural frequency of the system there is a rise in $\log_{10} AR$, in the underdamped system which is not apparent in the overdamped system. In the latter case the frequency response is similar in form to that of a first-order system (Fig. 4.9), except that the high frequency asymptote has a slope of -2 logs/decade. When the input frequency approaches the natural frequency of an underdamped system, the two become "in tune" with each other and the output amplitude increases. This phenomenon is called "resonance".

Thus, a periodic input applied to an underdamped system is most effective when its frequency approaches the natural frequency of the

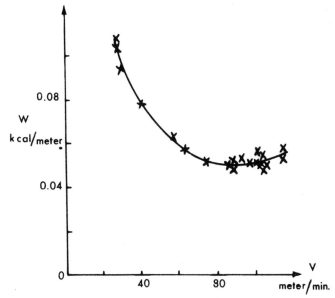

Fig. 4.12. Minimal energy expenditure occurs at a walking frequency (constant speed 3 m.p.h.) of about 80 m/min. (From Milsum, 1966; after Cotes and Meade, 1960.)

system. The best way to push a car out of a rut for example, is to "rock" it rhythmically, at the appropriate frequency. Similarly, less effort is required to attain a given output if a system is driven at its natural frequency. Again, most people have a preferred walking pace and claim that it is harder to walk faster or slower. Cotes and Meade (1960) determined the energy expenditure, measured by oxygen consumption, in human subjects walking at a constant speed, at various step frequencies. They found that energy expenditure was minimal at a frequency close to the natural frequency of the system (Fig. 4.12), as calculated from the length of leg, and other physical parameters.

4.2 THE PUPIL-CONTROL SYSTEM

The reflex regulation of pupil area has long been recognized as an example of feedback control and has recently become the subject of intensive study from the control theory viewpoint. The pupil has two major functions: regulation of the amount of light reaching the retina and assistance in maintaining image sharpness during near vision. In the latter, the pupil mechanism works in conjunction with the control systems for accommodation and convergence. The present discussion is concerned only with the former of these functions and is intended to illustrate some of the points raised in the foregoing part of this chapter.

4.2.1 Anatomical Considerations

In man the pupil is circular and its diameter is controlled by two antagonistic muscles, the *sphincter pupillae* and the *dilator pupillae.* The dilator muscle is innervated sympathetically via the superior cervical ganglion, and the sphincter muscle is parasympathetically innervated from the Edinger-Westphal nucleus, relaying in the ciliary ganglion. The sphincter muscle is of prime importance in the pupillary light-reflex and not only does stimulation cause pupil constriction, but dilation is probably caused by inhibition of the normal tonic condition.

Variation in pupil area causes changes in the illumination of the retina and consequent changes in messages transmitted via the optic nerve (Fig. 4.13). Some fibres, the pupillomotor fibres, go directly to the brain stem and relay in the pretectal nuclei from which the Edinger-Westphal nuclei are innervated. Thus a closed-loop is formed.

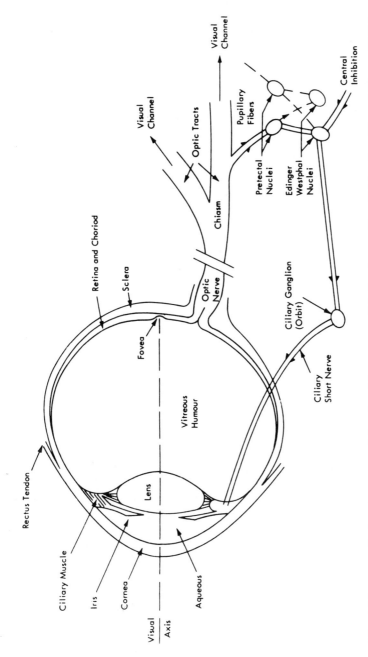

Fig. 4.13. Anatomical arrangement of the closed-loop responsible for the pupillary light-reflex. (From Milsum, 1966.)

The neural pathways serving the two eyes are cross-coupled. Illumination of one retina causes the consensual reflex, constriction of the pupil of the unstimulated eye as well as the stimulated eye. This factor, however, does not affect the closed-loop under consideration and such connections are omitted from the anatomical relations summarized in Fig. 4.13.

4.2.2 The Closed-loop

The closed-loop responsible for the pupillary light-reflex is summarized in Fig. 4.14. The retina is sensitive to the light flux F, in

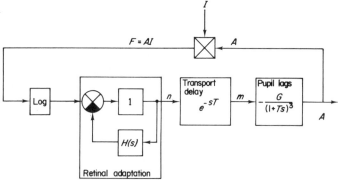

Fig. 4.14. Simplified block diagram of the closed-loop responsible for the pupillary light-reflex.

accordance with the Weber-Fechner law, so that for large signals the logarithm of the light flux is the operative variable. The retina also exhibits slow photochemical dark-adaptation and relatively fast light-adaptation. This retinal adaptation is partly responsible for the difference in transient response obtained when stimuli of opposite signs are applied (see section 4.2.3).

The delays involved in neural transmission within a single neurone are negligible with respect to the other delays in the system. The delays involved in synaptic transmission must be taken into account. A "pure time", "transport delay" (see Milsum, 1966, p. 214) is involved, such that

$$y(t) = x(t - T), \tag{4.12}$$

from eqn. (1.13)

$$y(s) = \int_0^\infty x(t - T) e^{-st} dt \tag{4.13}$$

$$y(s) = e^{-sT} \int_{T}^{T+} x(t-T) \, dt = e^{-sT}.$$ (4.14)

The delays involved in a series of synaptic transmissions can simply be added to give a single transport-delay (Fig. 4.14).

The sphincter muscle, responsible for changes in pupil area, is itself a complex sub-system involving a number of dynamic processes. According to Stark (1959) the relationship between pupil area A, and nerve signal m, can be simulated by the following equation,

$$Gm = T^3 \frac{d^3 a}{dt^3} + 3T^2 \frac{d^2 a}{dt^2} + 3T \frac{da}{dt}$$ (4.15)

where G is the gain of the pupil and T is the time-constant. The time-constant of constriction is shorter than that of dilation so that the representation of this sub-system as a third-order lag (Fig. 4.14) is an oversimplification.

The retinal flux F is the product of the pupil area A and the ambient light intensity I. In Fig. 4.14 this relationship is illustrated by means of a multiplier mixing point. The loop controlling the pupillary light-reflex is now closed and the resulting system can be regarded as a working hypothesis, to be examined experimentally.

4.2.3 Transient Analysis

Transient analysis of the pupillary light-reflex can be conducted by subjecting the eye to pulse, or step, changes in illumination, whilst measuring the 'pupil response. Changes in pupil area are measured by means of a pupillometer, by which infra-red light is reflected from the iris on to a photocell (Fig. 4.15). As more light is reflected from the iris than from the pupil, changes in pupil area have a marked effect on the amount of light reflected especially when the area of illumination is small.

Clynes (1961) used four types of stimulus in his transient analysis of the pupil response. The retina is adapted to a medium intensity of illumination and subjected to a brief (light-pulse), or sustained (light-step), increase in illumination; or to corresponding decreases in illumination (dark-pulse and dark-step) (see Fig. 4.16). Typically, a light-pulse produces a contraction and redilation of the pupil, and a dark-pulse produces a similar contraction, the reverse of the expected change. A light-step produces a contraction, which does not redilate

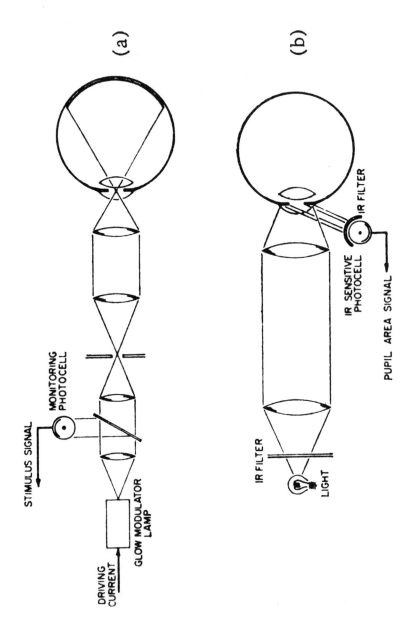

Fig. 4.15. Apparatus used in pupillary light-reflex studies: (a) servo-controlled light stimulation of the retina, (b) pupillo-meter by which infra-red light is reflected from the iris into a photocell. (From Stark, 1968.)

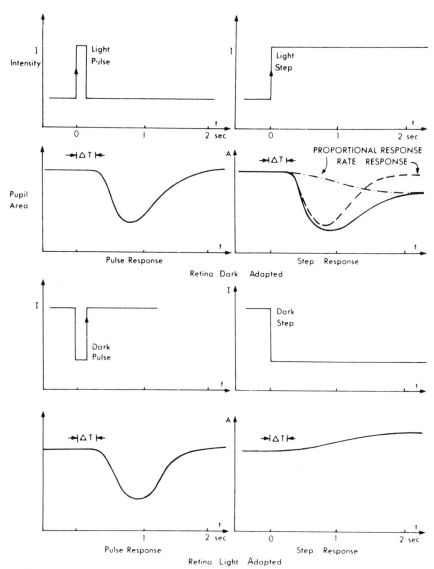

Fig. 4.16. Transient responses of the pupil to pulse and step inputs. (From Clynes, 1961.)

to the initial level, and a dark-step induces a slow dilation. These responses are illustrated in Fig. 4.16.

The observed phenomena can be accounted for in terms of the following three postulates: (1) A transport-delay of about 250 ms is responsible for the pause before any pupil response is detectable. (2)

A proportional response, which is sensitive to illumination changes in both directions, relates to the amount of light reaching the retina. (3) A rate response, sensitive only to increases in illumination, relates to the rate of change of illumination reaching the retina. The response to a light-pulse consists primarily of the rate response and the response to a light-step to the rate and proportional responses combined. However, the response to a dark-step lacks the rate sensitive component and consists solely of the proportional part. The response to a dark-pulse probably occurs solely to the reillumination component of the light stimulus.

Clynes (1961) accounted for the observed transient responses by means of the transfer function $H(S)$,

$$H(s) = \frac{be^{-T_s s}}{1 + T_4 s} + \Omega \frac{ase^{-T_3 s}}{(1 + T_1 s)(1 + T_2 s)} \qquad (4.16)$$

where Ω is an operator, which $= 0$, when

$$\frac{dx}{dt} < 0,$$

and otherwise $= 1$; a and b are sensitivity constants, or gain factors, and $T_1 - T_5$ are time-constants. The first term represents the proportional response of the pupil, and the second term represents the rate response to increases in retinal illumination. The latter is an example of the more general phenomenon of "unidirectional rate sensitivity" URS (Clynes, 1961). Unidirectional rate sensitivity is a common occurrence in biological systems, particularly where release of chemical substances is involved, because the release rate is generally much faster than the decay rate. URS can also result from the dynamics of muscular contraction, as in the case of the respiratory heart rate reflex (Clynes, 1961). It is probable that both chemical and muscular unidirectional phenomena are responsible for the URS observed in the pupillary light reflex. Unidirectional rate sensitivity can be responsible for a wide range of non-linear control phenomena which are discussed fully by Clynes (1961), Milsum (1966) and Milhorn (1966).

4.2.4 Frequency Analysis

Transient analysis of the pupillary light-reflex indicates that a non-linear control system is involved (section 4.2.3). The pupil-response exhibits a unidirectional rate sensitivity and in response to sinusoidal fluctuations in light intensity the pupil-response is non-sinusoidal (Clynes, 1961). Thus frequency analysis is not strictly applicable to this system. However, by using small sinusoidal inputs Stark and his co-workers were able to obtain a linear approximation to the pupil control system.

In the initial experiments (Stark and Sherman, 1957) the situation was simplified by breaking the feedback loop, as shown in Fig. 4.21. The sinusoidal input is applied in such a manner that changes in pupil area have no effect upon retinal illumination. Thus the system response has no influence over the input signal which remains entirely under the control of the experimenter. Under these conditions the transfer function is simplified, the "open-loop transfer function" $G(s)$ being related to the closed-loop transfer function $F(s)$ by

$$F(s) = \frac{G(s)}{1 + G(s)}. \qquad (4.17)$$

Studying systems under open-loop conditions, such as these, is a widely used method in engineering control systems analysis.

Experimentally breaking the feedback loop in a biological system

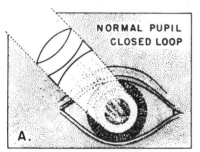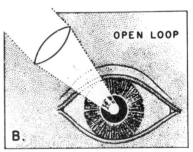

Fig. 4.17. Techniques used for stimulation of the pupillary light-reflex: (a) normal condition in which movement of the iris changes the light intensity at the retina, (b) light entering the pupil is unaffected by movement of the iris. (From Stark, 1968.)

is often a difficult problem, though in this case the solution is relatively simple. Instead of focusing the stimulus-light onto both pupil and iris, so that movements of the iris change the intensity at the retina (Fig. 4.17), the light is focused more acutely, so that it remains unaffected by changes in pupil area (Fig. 4.17).

For small sinusoidal inputs under open-loop conditions the pupil response is approximately sinusoidal, but also exhibits high frequency noise and low frequency drifting (Fig. 4.18). As a function

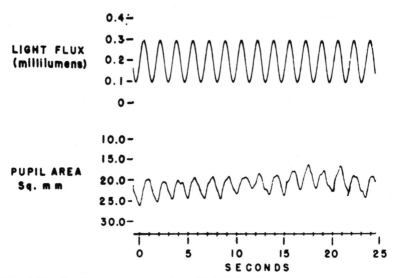

Fig. 4.18. Pupil response to sinusoidal light flux showing approximately sinusoidal response, plus high frequency noise and low frequency drift. (From Stark, 1968.)

of medium frequency the pupil response is complex. From the Bode diagram of Fig. 4.19 it can be seen that, from a low frequency gain of 0.16 the amplitude attenuates sharply to a high frequency asymptotic slope of −3 logs/decade, characteristic of a third-order system. On this basis, the phase shift at 4 cycles/s should be 270°, but is in fact twice this amount. When a system has the minimum phase-lag possible in relation to the observed AR plot it is said to have "minimum-phase" shift. When the phase-lag is greater than that indicated by the AR plot it is called a "non-minimum phase" shift. A transport delay has unity AR, but considerable phase shift, and is often responsible for non-minimum phase phenomena. In the present case, the non-minimum phase shift of 270° can be accounted for by a time delay of 0.18 s. On this basis, Stark (1959, 1968) gives the following estimation of the open-loop transfer function for the pupil control system

$$G(s) = \frac{0.16e^{-0.18s}}{(1+0.1s)^3} \tag{4.18}$$

From this transfer function, the low frequency closed-loop gain is calculated to be 0.14 which compares favourably with an experimentally determined closed-loop gain of 0.15 (Stark, 1959).

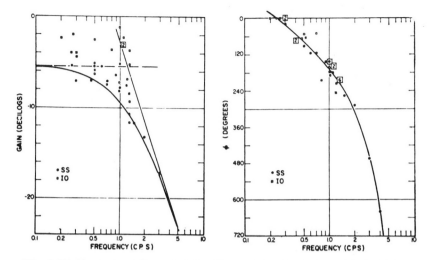

Fig. 4.19. Frequency characteristics illustrated as Bode plot of both high gain instability experiments (squares) (see section 4.2.4), and driven-response experiments (filled circles). (From Stark, 1968.)

The value of the transport delay is of the same order as that obtained from transient analysis. Thus, despite the non-linearities in the system, small-signal frequency analysis yields acceptable results.

It has been known for some time (Stern, 1944; Campbell and Whiteside, 1950) that a light beam focused on the edge of the pupil (Fig. 4.20) can induce pupillary oscillations. The effect of this procedure is to increase the gain of the system, since changes in pupil area have a larger than normal effect upon retinal illumination. A common result of increasing the gain of a closed-cloop system is to

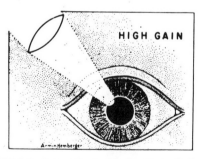

Fig. 4.20. When light is focused on the border of the iris and pupil, small movements of the iris result in large changes in light intensity at the retina. (From Stark, 1968.)

induce oscillations, particularly when a transport delay is involved (see Milsum, 1966, pp. 44-48).

Experimental manipulation of the system gain can be used as an analytical method. Stark (1962, 1968) manipulated the gain of the pupil control system by means of the procedure illustrated in Fig. 4.21. A break is made in the pupil closed-loop, by the method described above, and the output of the pupillometer used to drive an

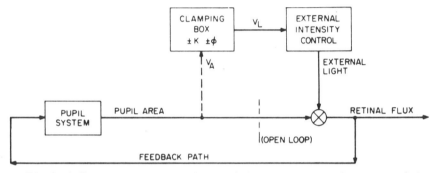

Fig. 4.21. Experimental manipulation of the system gain. The output of the pupillometer V_A is used to drive an electronic control network (clamping box), which in turn controls the stimulus light intensity. (From Stark, 1968.)

electronic control network which in turn controls the stimulus light intensity. So the natural loop is broken and then closed artificially. By manipulating parameters in the electronic control network the gain of the system can be altered experimentally and artificial phase lags can also be introduced. By suitable alteration of these parameters (for details see Stark, 1962, 1968), the pupil can be made to oscillate over a wide range of frequencies and a direct comparison made between Bode plots obtained from this, and from the standard, method of frequency analysis. Stark (1962) obtained good agreement between phase data obtained from the two methods but less good agreement for gain data (Fig. 4.19). The latter is probably due to the presence of a known non-linearity, logarithmic saturation which acts to reduce pupil gain for larger inputs.

4.2.5 Discussion

The pupil-control system affords an example of a quasi-behavioural system, which can be studied profitably both by means of transient analysis and by frequency analysis. Both methods have their advantages and limitations. Transient analysis shows up the non-linearities in the system which to some extent restrict the validity of frequency analysis. Nevertheless frequency analysis has proved a useful method, both in its own right and as a starting point for more advanced methods, to be described in a later chapter.

Frequency analysis has the particular advantage that signals of small amplitude can be used, avoiding some of the non-linearities, and enabling a linear approximation of the system transfer function. The work on the pupil response provides a good example of this procedure and also provides an example of the application in biology of a classical engineering technique, determination of the open-loop transfer function. However, the major lesson to be learnt from these studies is that biological systems are so complex, and contain so many special peculiarities, that only by using various different methods of analysis is an understanding of the mechanisms involved likely to be obtained.

CHAPTER 5

The Accuracy of Behavioural Control

It has long been claimed that domestic chicks become more accurate at pecking grain during the first few days after hatching. In investigating this phenomenon Hess (1956) fitted chicks with a latex hood arranged to carry prisms in front of the eyes, as shown in Fig. 5.1. The pecking response of birds fitted with prisms, which

Fig. 5.1. Chick fitted with latex hood designed to carry lenses and prisms. (From Marler and Hamilton, 1966; after Hess, 1956.)

displaced the binocular field seven degrees to the right, was compared with that of controls fitted with flat lenses. The birds were tested on the first or the fourth day after being hatched, and maintained in darkness. The results, illustrated in Fig. 5.2, indicate that both experimentals and controls, when aiming at a brass nail, deliver scattered pecks at one day old, while at four days pecks are

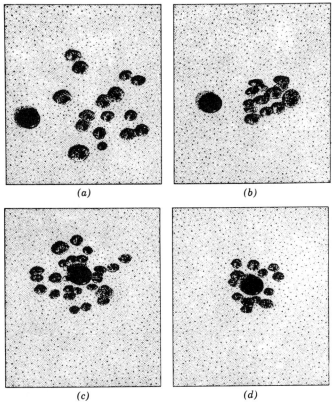

(a)   *(b)*

(c)   *(d)*

Fig. 5.2. Imprints of pecks aimed at a brass nail embedded in a soft surface. One-day-old (a) and four-day-old (b) chicks fitted with prisms show an unchanging constant error, in addition to the improvement in variable error shown by one-day-old (c) and four-day-old (d) chicks fitted with flat lenses. (From Marler and Hamilton, 1966; after Hess, 1956.)

grouped more together. This is an improvement of "variable error" and is, in this case, independent of experience and reward. The chicks fitted with prisms also showed a "constant error", to the right of the target, which did not improve with age, nor with subsequent experience. Clearly, the prisms provide a disturbance which is not compensated for by the control system involved. This distinction between variable and constant error must be taken into account in any consideration of accuracy in control systems.

5.1 CONSTANT ERRORS IN CONTROL SYSTEMS

5.1.1 Steady-state Errors

The mantid *Parastagmatoptera unipunctata* detects its prey visually, faces it by movement of the head and then strikes with a very rapid movement of the forelegs. The manner in which the mantid aims its strike in the horizontal plane has been extensively studied by Mittelstaedt (1957) and this work is discussed in section 2.3. Mittelstaedt found that the head-turning movement is controlled by optic and proprioceptive feedback, which enables the mantid to correct for disturbances due to changes in the position of the prey, and in the mechanical loading of the head (Fig. 2.13). However, the head does not come to face the prey but exhibits a constant error which Mittelstaedt calls the "fixation deficit". This constant error becomes apparent when the system reaches steady-state and is a typical example of a "steady-state error".

The manner in which the error arises can be seen from the block diagram of the system, portrayed in Fig. 5.3. In this diagram, ϕ

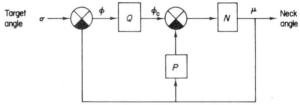

Fig. 5.3. Block diagram of system controlling prey-fixation in mantids (adapted from Fig. 2.13).

represents an error variable as the mantid would face the target if $\phi = 0$. The error variable does not reach this value because the system reaches steady-state before the error variable is completely reduced. At steady-state

$$\mu = \sigma \, \frac{QN}{1 + N(P + Q)} \tag{5.1}$$

$$\phi = \sigma \, \frac{1 + QN}{1 + N(P + Q)} . \tag{5.2}$$

Thus the magnitude of the steady-state error will depend upon the values of the parameters N, P and Q.

The steady-state error of the mantid facing a stationary prey is an error of "position". Other types of error can arise in tracking a target moving at a constant rate, when a system may show a "velocity" error. When the target is accelerating at a constant rate, an "acceleration" error can sometimes be observed. Position, velocity and acceleration errors can all be regarded as steady-state errors when exhibited by systems subject to inputs which are constant functions of time.

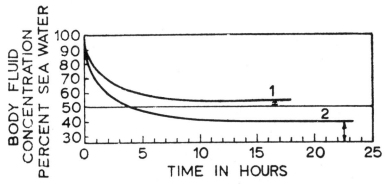

Fig. 5.4. Steady-state error, indicated by arrows, of polychaete worm *Nereis diversicolor*, when immersed in 50% seawater (1) and 25% seawater (2). (From Bayliss, 1966.)

Consider, as an example, the marine polychaete worm *Nereis diversicolor*, which has a simple form of osmotic regulation by means of which the concentration of the body fluids is maintained fairly constant despite fluctuations in the external medium. If suddenly transferred from sea water to 50% sea water, the body fluids become equivalent to 55% sea water, as shown in Fig. 5.4. When transferred to 25% sea water, the body fluids become equivalent to 40% sea water (Bayliss, 1966). Thus the response to a step-function in external concentration is typical of a first-order system exhibiting a position error. An animal with a more sophisticated form of osmotic regulation would probably not show such a position error but might well show a velocity error, as it "tried to keep up", when subjected to an external medium in which the concentration was changing at a constant rate.

The types of steady-state error exhibited by a control system depend jointly upon the system dynamics and the type of input to which the system is subjected. The response to standard position, velocity and acceleration inputs, such as those illustrated in Fig. 5.5, is often used as a means of classifying control systems (see Milhorn, 1966; Sensicle, 1968). This procedure leads to mathematical considerations which are of prime importance in the design of control systems.

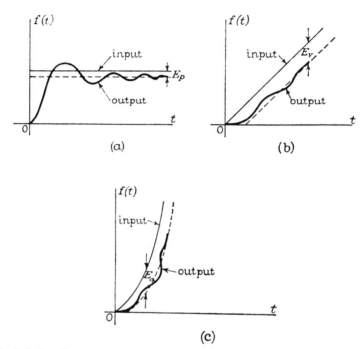

Fig. 5.5. Steady-state errors, E_p, E_v, E_a, in response to position (a), velocity (b) and acceleration (c) inputs, respectively. (From Sensicle, 1968.)

5.1.2 Compensation

Steady-state errors, and other "undesirable" performance characteristics, can often by compensated for in the design of control systems by the introduction of appropriate sub-systems which counteract the cause of the error. Consider for instance the simple feedback system illustrated in Fig. 5.6. The transfer function is

$$\frac{Y}{X}(s) = \frac{a}{s(s+ab)} \tag{5.3}$$

from which $Y(t) = \alpha \frac{a}{ab} (1 - e^{-abt}). \tag{5.4}$

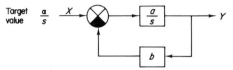

Fig. 5.6. Simple first-order feedback control system.

Thus the asymptotic output α/b fails to meet the target value α and the system exhibits a steady-state error (Fig. 5.7). The error can be reduced by introducing a simple sub-system onto the input side of the comparator, as shown in Fig. 5.8. In this case

$$\frac{X}{Y}(s) = \frac{ab'}{s(s+ab)} \tag{5.5}$$

$$\text{and} \quad Y(t) = \alpha\frac{ab'}{ab}(1-e^{-abt}). \tag{5.6}$$

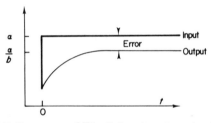

Fig. 5.7. Response of Fig. 5.6 system to a step input.

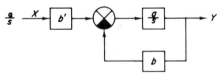

Fig. 5.8. Compensation of Fig. 5.6 system by introduction of input parameter b'.

When the transfer function of the introduced sub-system is equal to that of the feedback path, i.e. $b' = b$, the asymptotic output equals the target value α (eqn. 5.6).

Correction of performance errors by such means is common practice in engineering control systems design and is generally called "compensation". Machin (1964) suggests that systems which derive their input from the CNS may correct for the error-producing characteristics of feedback pathways, by means of compensation provided centrally. Compensation can take many forms depending upon the nature of the control system and the performance criteria required of it. A common method is to alter the nature of the error-actuated controller. If the controlled system exhibits undue lag due to the action of integrators, then a compensating lead can be introduced by incorporating differentiators into the controller, as shown in Fig. 5.9. Many means of compensation of this basic type are discussed by Milsum (1966). Another form of compensation is to

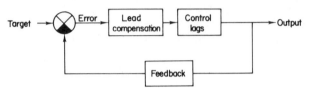

Fig. 5.9. Control lag can sometimes be eliminated by introduction of lead compensation in the form of differentiating elements.

"anticipate" the consequences of feedback loops by means of "feed-forward" control. The basic principle of feed-forward control is illustrated in Fig. 5.10. Most mammals control body temperature on the basis of information received from temperature receptors in the hypothalamus. Humans also have well-developed peripheral receptors by means of which information concerning changes in ambient temperature can be fed directly to the CNS, bypassing the thermal lags involved in central reception (Benzinger, 1969). Essentially, the peripheral receptors "anticipate" changes in body temperature of environmental origin.

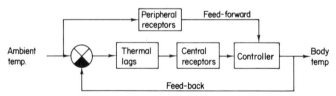

Fig. 5.10. Simplified representation of feed-forward control provided by peripheral thermoreceptors.

Feed-forward is a phenomenon which enables an animal to anticipate the long-term consequences of behaviour and to take appropriate action to forestall such consequences. This principle is particularly important for the activities that contribute to the maintenance of homeostasis (McFarland, 1970a). For example, the long-term digestive consequences of food intake generally increase thirst, but instead of drinking as a result of such a thirst increase, many animals drink in advance and thus anticipate the effects of food ingestion. A similar phenomenon can be demonstrated with respect to the effect of ambient temperature upon drinking and examples of this type are discussed in Chapter 4.

5.1.3 Performance Criteria

The notion that inaccuracies can be compensated for by incorporating certain features into the design of a control system implies that performance be assessed in relation to definite criteria. Performance has so far been discussed in terms of error-variables in error-actuated systems and the reduction of such errors to zero, as fast as possible, provides a simple form of "performance criterion". However, it is often desirable to take a variety of criteria into account, particularly those relating to the "cost" of the control operation. In the case of walking, discussed in section 4.1.4, it is clear that the system is designed to minimize energy expenditure as a function of walking speed. The "cost" is less when the step-frequency is close to the resonant frequency of the system.

In the case of the response of a second-order system to a step-input, illustrated in Fig. 5.11, three commonly used per-

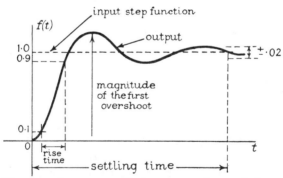

Fig. 5.11. Three performance criteria commonly used in assessing the response of second-order systems. (From Sensicle, 1968.)

formance criteria are indicated. The "rise time" is the time taken for the output to rise from 10% to 90% of the target value. The magnitude of the first overshoot provides another criterion, as does the "settling time", which is the time taken for the output to reach and remain within 2% of its final steady-state value. A situation in which all three of these criteria are likely to be of importance is the bringing of a motor boat alongside a landing stage (Fig. 5.12). Given

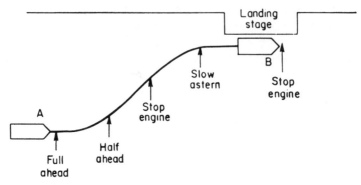

Fig. 5.12. Method of handling a motor boat to bring it alongside a landing stage. (From Wilkins, 1966.)

continuous control over the motor speed a suitable performance on the part of a human operator would be that illustrated in Fig. 5.13. The knack is to apply judicious underdamping to overcome the inertia of the boat.

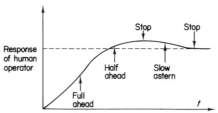

Fig. 5.13. Response of human operator, in terms of position of engine control lever, in bringing a boat alongside a landing stage.

A more concrete example is this simple experiment, that can be satisfactorily conducted in the undergraduate laboratory. The apparatus consists of a thermal system, constructed from a beaker and a test-tube (Fig. 5.14). The beaker is filled with water which can be heated by means of a manually controlled electrical heater. The test-tube, also filled with water, is suspended inside the beaker and the temperature of the water in the test-tube is measured by means of a suitable thermometer. The voltage applied to the heater and the thermometer reading should be recorded as a function of time.

The human subject of the experiment is given the task of heating the water in the test-tube so that it reaches and remains at a specified target temperature within the shortest possible time. Thus the performance of the subject is clearly defined in terms of a settling-time criterion.

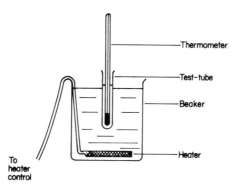

Fig. 5.14. A simple thermal system which can be used to illustrate human performance in a feedback control task.

Before considering the subject's performance it is necessary to determine the dynamic characteristics of the apparatus. When a voltage is applied to the heater, the temperature of the beaker rises exponentially in accordance with the laws of heat transfer. This process can be represented in block diagram form (Fig. 5.15) as a first-order thermal lag (Milsum, 1966). Similarly, the heat transfer,

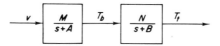

Fig. 5.15. Simplified block diagram of the thermal properties of the apparatus illustrated in Fig. 5.14. v = voltage supplied to the heater, T_b = temperature of water in the beaker, T_t = temperature of water in the test-tube, M, N, A and B are thermal parameters.

from the water in the beaker to that in the test-tube, can also be represented as a first-order lag (Fig. 5.15). When a step-increase in voltage ($V = {}^v\!/s$) is applied to the system, the temperature response T_t will be determined by the two lags in cascade, viz.

$$T_t(s) = \frac{v}{s} \cdot \frac{M}{(s+A)} \cdot \frac{N}{(s+B)} \qquad (5.7)$$

$$T_t(t) = v\frac{MN}{AB} \left(1 - \frac{Be^{-At}}{B-A} - \frac{Ae^{-Bt}}{B-A}\right). \qquad (5.8)$$

If the human operator turned the voltage, step-wise, to the correct final steady-state level, the temperature would rise slowly to the target value in accordance with eqn. (5.8), as shown in Fig. 5.17. This would be the best performance that could be obtained by simple open-loop control. The subject can however, visually compare the temperature at any time with the target temperature, indicated on the thermometer. The error can be used to determine the input voltage, so that a closed-loop is formed (Fig. 5.16). Obviously, the performance of the system under these conditions will depend upon the transfer function of the human operator.

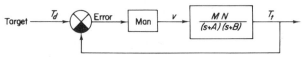

Fig. 5.16. Closed-loop formed when a human utilizes visual comparison of target and actual temperature to determine the voltage applied to the heater.

Suppose the response of the human subject were proportional to the error, so that the function Man in Fig. 5.16 is replaced by the parameter K. Then the overall transfer function of the system,

$$T_t/T_d(s) = \frac{KMN}{(s+A)(s+B) + KMN}$$

$$= \frac{KMN}{s^2 + s(A+B) + AB + KMN},$$ (5.9)

would be typical of a second-order system, such as that discussed in section 4.1.4. Thus we can specify the following characteristics

steady-state asymptote $\quad A = \dfrac{KMN}{AB + KMN}$ (5.10)

natural frequency $\quad \omega_n = \sqrt{AB + KMN}$ (5.11)

damping ratio $\quad \zeta = \dfrac{A+B}{2\sqrt{AB + KMN}}.$ (5.12)

When K is large such a system is likely to give an underdamped response, like that illustrated in Fig. 5.17. But the performance of the system, in relation to the settling-time criterion, can be improved by various types of compensation. In particular, if the human operator uses derivative plus proportional control, instead of proportional control only, a considerable improvement in performance is achieved. Such control implies that the human subject is sensitive to the rate-of-change of error, in addition to its actual value. Thus the transfer function $K(s)$ is replaced by $(K+s)$, and the overall transfer function of the system is then as follows

$$T_t/T_d(s) = \frac{MN(K+s)}{MN(K+s)+(s+A)(s+B)} \tag{5.13}$$

steady-state asymptote $A = \dfrac{KMN}{AB+KMN}$ $\qquad$ (5.14)

natural frequency $\qquad \omega_n = \sqrt{AB+KMN}$ $\qquad$ (5.15)

damping ratio $\qquad \zeta = \dfrac{A+B+MN}{2\sqrt{AB+KMN}}$ $\qquad$ (5.16)

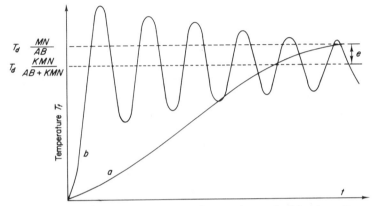

Fig. 5.17. The temperature response is overdamped: (a) when the voltage input is pre-set to the correct final value, and exhibits underdamping, (b) plus steady-state error (e), when proportional feedback-control is used.

Here the damping ratio increases, while the natural frequency remains the same as does the steady-state asymptote. Thus there is an improvement in the rise-time of the response but the steady-state error remains (Fig. 5.18). The performance can be further improved by the addition of error-integral compensation, so that the total control consists of proportional plus derivative plus integral components (Fig. 5.18). The integral control acts as a follow-up device which continues to increase the gain as long as an error exists. Thus the effect of adding error-integral compensation is to remove the steady-state error (see Milsum, 1966; Sensicle, 1968) and, when combined with derivative-compensation, the performance is greatly improved (Fig. 5.18).

The combination of proportional plus derivative plus integral control, often called "three-term control", is frequently employed by design engineers. It is probably common in biological systems also, for Goldman (1961), discussing the glucose-control system,

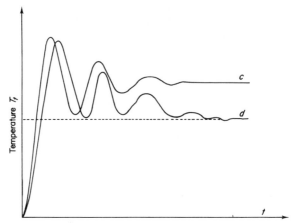

Fig. 5.18. Underdamping of the response is improved by the use of proportional-plus-derivative control (c) and the steady-state error removed by the addition of integral control (d).

suggests that all three forms of control are present and that fat-deposit storage provides the integral mode.

It appears that the human operator may initially use only proportional control in the temperature control task outlined above. With increasing experience the performance improves (see Fig. 5.19) and it is probable that both derivative and integral control are brought into play (Cooke, 1965). As the human operator becomes more skilled, feedback control is dispensed with and a sophisticated form of predictive control is used (see Fig. 5.20).

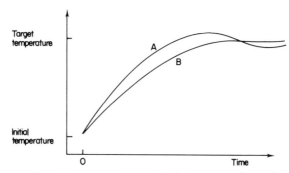

Fig. 5.19. Typical performances obtained during an undergraduate practical class. On the second trial (*A*) the subject shows a tendency towards underdamping, which has disappeared by the fifth trial (*B*).

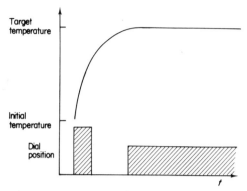

Fig. 5.20. Performance on the eighth trial is greatly improved, caused by the use of "bang-bang" open-loop control in which the control dial is turned fully up for a carefully judged period, then fully down for a further period, and then exactly to the position required to maintain a steady target temperature.

Empirical evidence suggests that the simpler the task imposed upon the operator, the more precise and less variable become his responses. Birmingham and Taylor (1954) suggest that optimal man-machine performance can be obtained when the mechanical components are so designed that the human need only act as a simple amplifier. Figures 5.21 and 5.22 indicate how this can be achieved, by shifting the burden of compensation from the human to the mechanical part of the system.

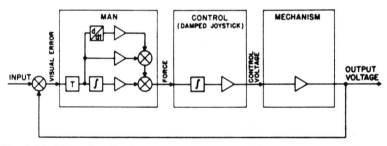

Fig. 5.21. Man, visualized as a three-term controller in a tracking system with damped joystick control and no aiding. T/T represents the delay induced by the human reaction time. (From Birmingham and Taylor, 1954.)

5.1.4 Adaptive Control

The human operator in a control situation is able to adapt his method or strategy of control as a result of experience, or of

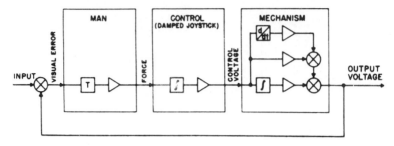

Fig. 5.22. When aiding is provided by incorporation of a three-term controller into the mechanism, man need only act as a simple amplifier. (From Birmingham and Taylor, 1954.)

changing operational requirements. In the steady-state, the operator adjusts his own characteristics (Krendel and McRuer, 1960) and is also able to adapt to gradual or sudden changes in the dynamics of the controlled mechanism (Young *et al.,* 1963). In the temperature-control experiment, discussed above, the subject rapidly improves his performance with experience and moves from closed-loop to a sophisticated form of open-loop control (Cooke, 1965). The attainment of such skilled performance is readily demonstrable in the laboratory (Fig. 5.20).

Improvement of performance, by changing the properties of the control system, is generally termed "adaptive control". This ability is not confined to humans but occurs throughout the animal kingdom, at many levels of organization. Adaptive control is a topic of great current interest in both physical and biological control systems analysis, though the mechanisms involved are far from understood.

Consider the drinking-control system outlined in section 3.3.2. Satiation of drinking ultimately depends upon absorption of ingested water from the lumen of the intestine. The delay involved in the absorption process may be such that drinking is terminated before there can be any appreciable change in the state of the blood. Oral and alimentary factors are usually invoked to account for rapid satiation phenomena. However, experiments with oesophageal fistula indicate that oral factors, though essential for reinforcement, play no part in satiation under the experimental conditions relevant to the present study (McFarland, 1969b). As for satiation, the system (Fig. 5.23) is characterized by the following transfer function (see eqn. 3.12).

$$I/L(s) = \frac{k(s+h)}{s^2 + s(h + kg) + hfk} \tag{5.17}$$

Thus

$$\omega_n = \sqrt{hfk} \tag{5.18}$$

and $\quad \zeta = \dfrac{h + kg}{2\sqrt{hfk}} \tag{5.19}$

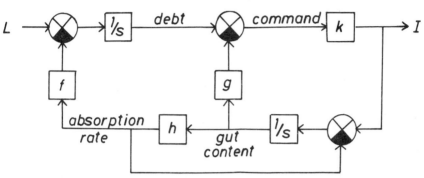

Fig. 5.23. Simplified system for satiation of operant drinking behaviour. (From McFarland and McFarland, 1968.)

The physiological implication of these equations is to be seen in the nature of the parameters of the system. The absorption constant h determines the time-constant of the absorption delay. The parameter f represents the "hydrating power" of the absorbed water and can be taken to vary with the salinity of the drinking water. So saline would be a less effective hydrating agent than distilled water. The effectiveness of the short-term satiation feedback is determined by the parameter g, which represents the combined effect of the sense organs involved and of the calibration of the message in the brain. The parameter k represents the gain of the ingestion mechanism.

It is reasonable to assume that, in this case, the relevant performance criterion is that the system should be critically damped. Under such conditions ultimate satiation is achieved as quickly as possible, without causing systemic overhydration. To maintain critical damping it is necessary that the value of g be calibrated to match the salinity of the water that is generally available. When $f = g$ eqn. (5.17) factorizes. Thus let $f = g = j$, then from eqn. (5.17)

$$I/L(s) = \frac{k(s+h)}{(s+h)(s+kj)} = \frac{k}{s+kj}. \tag{5.20}$$

In response to a step-function in the water debt, such as occurs when water is presented after a period of deprivation, the rate of water intake would decline exponentially to its normal level and the amount of water drunk during the first drinking bout of recovery from deprivation is given by the following equation

$$I/s = \frac{Dk}{s(s+kj)}, \mathcal{L}^{-1} I/s = \int Idt = D/j(1 - e^{-kjt}) \tag{5.21}$$

where D is the value of the water debt.

In an operant situation, where the gain of the ingestion mechanism is artificially reduced, this satiation curve can be empirically confirmed (see Fig. 3.22). When the value of g is reduced to match a reduced f the time-constant of the satiation curve is increased and as the animal gains experience with saline, after being trained with distilled water, this change should become apparent. The experiment described below was designed as an empirical test of this prediction.

Four adult Barbary doves *Streptopelia risoria* were trained to peck at an illuminated key in a Skinner box. Each peck was followed by a 2-s extinction of the key light, during which futher responses were inoperative, and 0.1 cc water was delivered to a small cup from which the subject drank. This reward procedure has previously been found to result in an approximately exponential satiation curve (McFarland and McFarland, 1968). After training, the subjects were housed in visual isolation at a controlled temperature (20°C) and artificial daylength (8 h light; 16 h dark). Food was provided *ad libitum* but water was available only during test sessions. Each subject was tested at the same time each day and run in the apparatus until it achieved a satiation criterion of 5 min without responding. An event recorder was used to make a continuous record of responses as a function of time.

After training, the subjects were tested for alternating periods of seven days using distilled water or 0.5% saline as a reward. This concentration of saline is below the rejection threshold for pigeons (Duncan, 1960) and is not discriminated from distilled water by doves. The experiments were conducted on a Latin square basis, so that each subject acted as its own control.

The results are expressed in terms of the time-constant of the satiation curves recorded during each daily test session, which are estimated from the slope of the logarithm of the integrated response rate plotted against time. These data are summarized in Fig. 5.24, and an example from a single subject is illustrated in Fig. 5.25. The results show that, on the first day of change from water to saline reward, the time-constant of the satiation curve is increased, as found in previous work (McFarland and McFarland, 1968). In addition,

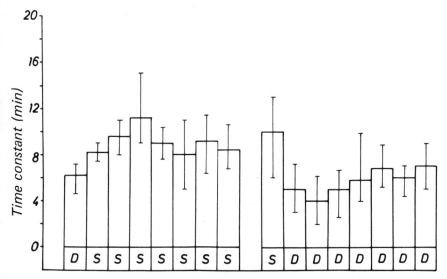

Fig. 5.24. Time-constants of satiation curves obtained in successive tests with saline S or distilled water D. Vertical lines indicate the range.

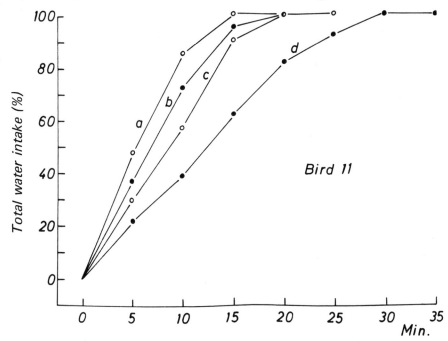

Fig. 5.25. Satiation curves of an individual subject in four successive tests with distilled water (a) and saline (b, c, d).

there is a progressive increase over the next two days, as predicted from the theoretical considerations outlined above. After the third day with saline reward the time-constant is reduced. This unexpected result may be due to an adaptive change in the mechanisms responsible for salt excretion. Such a change would effectively correspond to a change in the hydration factor f and a reduction in time-constant would be understandable on this basis.

When reward is changed from saline to distilled water, there is a marked reduction in the satiation time-constant (Fig. 5.24), as would be expected if the system suddenly became underdamped. In other words, an animal used to working for saline takes less distilled water as the effective absorption is enhanced with respect to the short-term satiation mechanism. That the time-constant gradually returns to its normal value as water rewards are continued, suggests that recalibration of alimentary signals also occurs under these conditions.

The shifts in the satiation time-constant observed when reward is changed from saline to water, and vice versa, are consistent with predictions made on the assumption that optimum regulation is achieved when the satiation feedback system is critically damped. It is envisaged that critical damping is maintained by calibration of alimentary signals indicating the quantity of water in the alimentary tract. It is possible that the calibration is made, within the CNS, on the basis of learned post-absorptive changes in water balance. Recent work (Garcia et al., 1966, 1967) indicates that conditioning can occur through physiological after-effects many hours after eating, and it is reasonable that similar principles should apply in the present case. The implication of this view is that regulation is achieved by means of an adaptive control system which acts to preserve an optimum state such that the amount of water ingested is appropriate to the needs of the animal. Unless the system is critically damped the animal is liable to overshoot or undershoot the mark.

Adaptive control may involve much more than simple adjustment of parameter values in closed-loop control systems (Gibbs, 1970). For example, the work of Denier van der Gon and Wieneke (1969) suggests that the stretch reflex is suppressed during fast hand movements, so that the control system involved is closed-loop or open-loop, depending upon the use made of the neuromuscular system. It appears that many skilled movements, such as handwriting (Denier van der Gon and Thuring, 1965), are controlled directly by "preprogrammed" signals generated in the brain and that no feedback is involved. The old Chinese bamboo painters trained themselves to use the brush in one movement as retouching was

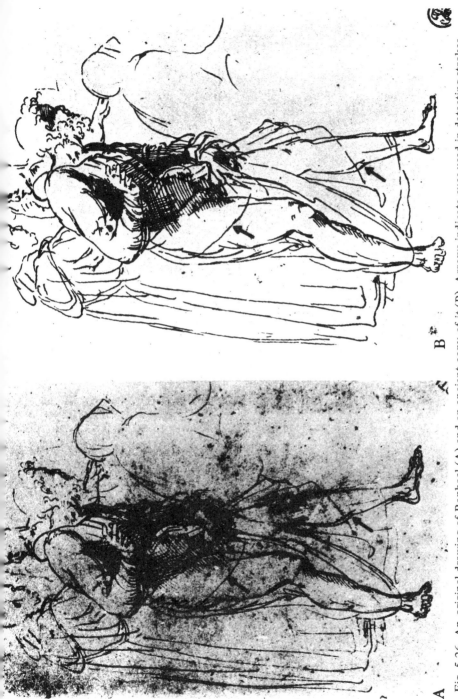

Fig. 5.26. An original drawing of Raphael (A) and an ancient copy of it (B). Arrows indicate cues used in detecting strokes made under visual guidance, which the original artist had executed without direct visual feedback. (From van der Tweel, 1969.)

impossible in their material. This type of difference between the master and the less-skilled artist is employed in the detection of forgeries (van der Tweel, 1969). The master executes the pen or brush stroke confidently and swiftly, but the copyist is not always able to follow his example, as indicated in Fig. 5.26.

5.2 VARIABLE ERRORS IN CONTROL SYSTEMS

5.2.1 Qualitative Aspects

Variability in behavioural data is a phenomenon frequently encountered by research workers. Most commonly it is due to failure to control for random fluctuations in disturbance variables. In some cases variability persists even when careful controls are incorporated into the experimental design. This had led some workers to assume that behavioural systems are by their very nature stochastic processes (Thurstone, 1919; Miller and Frick, 1949; Estes and Burke, 1953). An alternative view is that behavioural systems are basically deterministic, and that the problem of variability should be met by attempts to identify the source of variation and then to specify its cause (Chapanis, 1951).

At the purely qualitative level, it is sometimes possible to identify sources of variation by simple experimental means. For example, day-to-day fluctuations in food and water intake commonly occur in animals maintained under laboratory conditions. McFarland (1967) observed that in doves, maintained under constant temperature conditions, the variations in food and water intake tend to be in phase with each other (Fig. 5.27) but that this is not so when ambient temperature is not controlled. As food intake largely dictates water intake under *ad libitum* conditions (Bolles, 1961; McFarland, 1965a; Oatley and Tonge, 1969), it is likely that the in-phase variations, observed under constant-temperature conditions have their origin in the feeding system. On the other hand, variations in the drinking system would be expected to cause out-of-phase variations in feeding and drinking as a result of the inhibition of food intake by thirst (McFarland, 1964; Oatley, 1967). Such out-of-phase changes can be experimentally induced by changes in environmental temperature (Fig. 5.28), or by water deprivation (McFarland, 1967). Thus variations in feeding and drinking can be interpreted in terms of known interactions between the feeding and drinking systems. Feeding and drinking vary directly when the feeding system is the

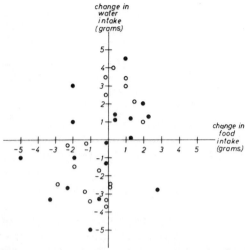

Fig. 5.27. Magnitude of day-to-day changes in water intake as a function of corresponding data for food intake, during constant temperature conditions. (Open and filled circles denote different subjects.) Under variable temperature conditions the numbers of points in each quadrant are approximately equal. (From McFarland, 1967.)

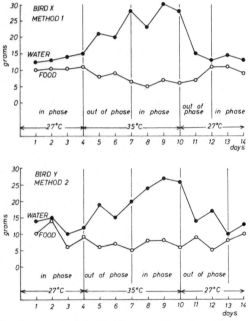

Fig. 5.28. Daily food and water intake as a function of temperature change. (From McFarland, 1967.)

main controlling factor, but vary inversely when the drinking system is dominant in the interaction between hunger and thirst.

The possible sources of behavioural variability are numerous. Random fluctuations in environmental variables, such as temperature, visual stimuli, etc. are particularly important in the wild. In the laboratory steps are usually taken to suppress, or control, such disturbances. Variable errors can arise in sensory processes themselves, either as a result of "noisy" monitoring mechanisms, or in the process of signal detection at higher levels in the nervous system (Green and Swets, 1966). Some workers (e.g. Dawkins, 1969a; Gray and Smith, 1969) have sought to account for behavioural variability in animals in terms of decisions made in the face of randomly fluctuating "drive". Similar assumptions are often involved in models of human choice behaviour (e.g. Luce, 1959; Audley, 1960). Variability in performance may also be due to errors in computation (Rabbitt, 1968), to variability of feedback from consequences of behaviour (Notterman and Mintz, 1965), or to motor processes (Harter and White, 1968). Distinctions between the various possible sources of noise, at a reasonably sophisticated level, must be made experimentally on the basis of quantitative measurement.

5.2.2 Measurement of Variability

The degree to which repeated behavioural measurements exhibit variability depends partly upon the unit of measurement employed. When the unit of measurement has been determined on the basis of criteria independent of variability (section 3.1.1), the investigator must take the variability into account in interpreting the data.

A set of data can be typified, or represented, by measures of central tendency or "average". Several types of average can be defined and when the data are obtained from discrete observations measurements of mean, median, or mode are employed, in accordance with the nature of the data and the purpose of the analysis (Moroney, 1951; Spiegel, 1961). When the observations form a time series, the "mean" can be defined as follows

$$\mu_x = \lim_{T \to \infty} \frac{1}{T} \int_0^T x(t)dt. \qquad (5.22)$$

The mean square value about the mean, called the "variance", is

$$\sigma_x^2 = \lim_{T \to \infty} \frac{1}{T} \int_0^T (x(t) - \mu_x)^2 \, dt \tag{5.23}$$

and its positive square root is the "standard deviation", σ_x.

In cases where average values are of little interest, the general intensity of random data can be described by the "mean square value"

$$\psi_x^2 = \lim_{T \to \infty} \frac{1}{T} \int_0^T x^2(t) \, dt \tag{5.24}$$

and the positive root of this is called the root mean square, or "rms" value. The variance is equal to the mean square value minus the square of the mean, viz.

$$\sigma_x^2 = \psi_x^2 - \mu_x^2. \tag{5.25}$$

A random process is said to be "stationary" when its statistical properties do not change with time. In such cases it is often useful to record only samples of the time-history and to relate the statistical properties of the data to the total "ensemble". If the random process is stationary, and the ensemble-average and other statistical processes do not differ when computed over different sample functions, the process is said to be "ergodic". Considerable progress has been made in the analysis of stationary, ergodic, random processes but relatively little has been achieved in the analysis of non-stationary phenomena. For details of the methods employed in random process analysis the reader is advised to consult the relevant literature as only the bare outline is presented here. An introduction to the subject is contained in Brown (1965) and Milsum (1966). More definitive accounts are given by Bendat and Piersol (1966), Cox and Lewis (1966) and Lee (1960).

In addition to specifying the mean and variance of a set of data, it is often convenient to determine the distribution of the data in terms of some variable of interest. Thus the number of events, or of intervals between events, may be plotted against their amplitude in the form of a histogram (Fig. 5.29). The probability of occurrence of events of given amplitudes is obtained by dividing the number of events $n(x_k)$ by the total number of events N, viz.

$$P(x_k) = \frac{n(x_k)}{N}, \qquad \sum_{k=1}^{K} P(x_k) = 1 \tag{5.26}$$

The probability that the data will assume a value within some

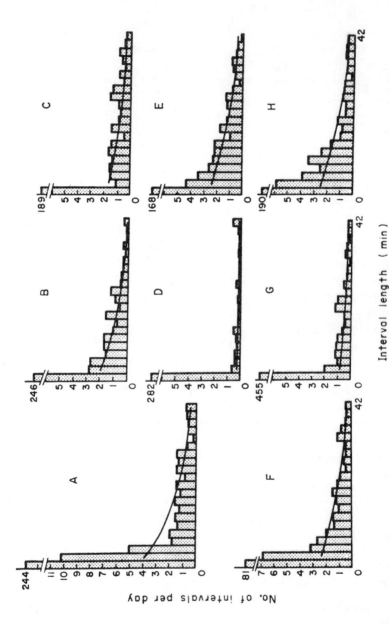

Fig. 5.29. Frequency histogram of intervals between feeding bouts recorded from eight chickens. The curves represent the theoretical distribution calculated on the basis that feeding is occurring at random for all intervals above 2 min. (From Duncan et al., 1970.)

specified range during a particular time-interval is given by the "probability density function"

$$p(x) = \frac{dP(x)}{dx} \qquad (5.27)$$

where the probability of the value of the variable falling between x and $(x + \delta x)$ is given by the expression

$$P(x + \delta x) - P(x) = x \cdot \frac{d[P(x)]}{dx} = \delta x \cdot p(x). \qquad (5.28)$$

As the value of the signal must lie within the range $-\infty < x < \infty$, it follows that

$$\int_{-\infty}^{+\infty} p(x)dx = 1. \qquad (5.29)$$

The probability that a value between x and $(x + \sigma x)$ will be observed is equal to the fraction of time spent by the signal between these amplitude limits. Thus, let T_x be the total amount of time that $x(t)$ falls inside the range x to $(x + \sigma x)$ during each observation period T (Fig. 5.30). As T approaches infinity,

$$p[x < x(t) \leqslant (x + \delta x)] = \lim_{T \to \infty} \frac{T_x}{T} \qquad (5.30)$$

The probability density function

$$p(x) = \lim_{\delta x \to 0} P \frac{[x < x(t) < (x + \delta x)]}{\delta x} = \lim_{\delta x \to 0} \lim_{T \to \infty} \frac{1}{T} \left(\frac{T_x}{x} \right). \qquad (5.31)$$

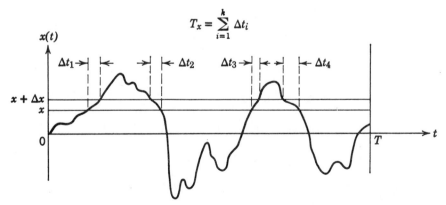

Fig. 5.30. Probability measurement. (From Bendat and Piersol, 1966.)

Figure 5.31 illustrates four examples of time-histories, together with their probability density functions. For convenience the mean value is assumed to be zero in each case. The probability density functions of the signals having a periodic component often show a bimodal form, while those for purely random data are typically Gaussian. These examples show how the probability density function

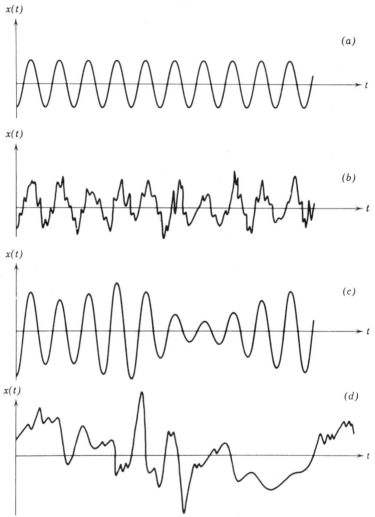

Fig. 5.31. Four time histories and their probability density functions. (a) Sine wave, (b) sine wave plus random noise, (c) narrow-band random noise, (d) wide-band random noise. (From Bendat and Piersol, 1966.)

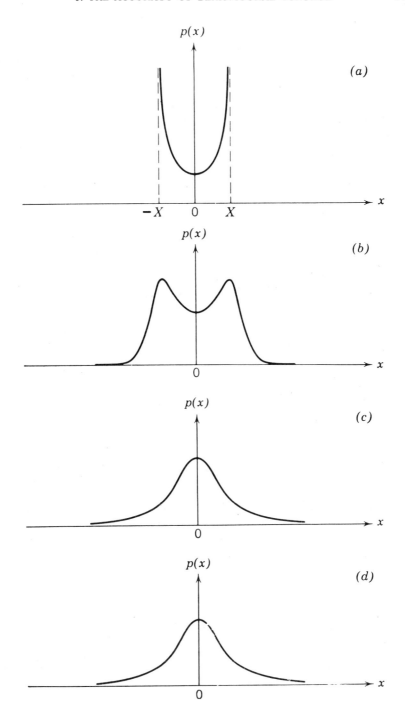

can be of assistance in determining the nature of variability in data obtained during experimental systems analysis. In particular the Gaussian and Poisson distributions, which imply that independent processes are at work in producing the observed variation, can be applied in time-series analysis (Bendat and Piersol, 1966; Cox and Lewis, 1966) as in ordinary statistics (Moroney, 1951; Spiegel, 1961). The Gaussian function also has important theoretical applications in systems analysis (Milsum, 1966).

5.2.3 Correlation Functions

The variability of behaviour with time poses one of the most formidable problems in the interpretation of behaviour patterns in terms of the underlying organization and integration of behavioural components. A common approach is to determine transition frequencies between one activity and another. Baerends *et al.* (1955) observed sequences of courtship activities in the male guppy *Lebistes reticulatus*. They calculated the transition frequencies between the various components of the courtship behaviour and indicated these in diagrammatic form, as shown in Fig. 5.32. As a further step, data of this type can be subjected to factor analysis (Wiepkema, 1961), or used to construct "grammatical" rules (Vowles, 1970; Hutt and

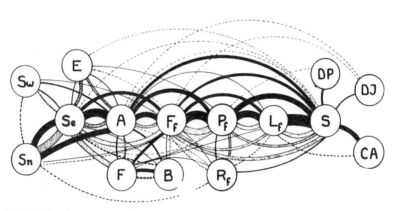

Fig. 5.32. Sequence of courting activities shown by male guppies. The thickness of the lines indicates the frequency with which one activity is followed by the one to which the line leads. Black lines should be read from left to right, hatched bands and dotted lines, from right to left. A = approach, B = biting, CA = copulation attempt, DJ = display jump, DP = display posturing, E = evading, F and F_f = following, L_f = luring, P_f = posturing, R_f = retreating, S = sigmoid posture, Se = searching, Sn = snapping, Sw = swimming about. (From Baerends, Brouwer and Waterbolk, 1955.)

Hutt, 1970). Hinde (1970) points to a number of difficulties in interpreting the results of studies involving sequential correlations. Although it is tempting to assume that two activities are likely to share common causal factors if they are closely associated in time, many such associations are liable to be due to changes in the external stimulus situation consequent upon the animal's behaviour. Further, the evidence suggests that causally unrelated activities may appear at particular points in a behaviour sequence as a result of motivational equilibrium, or other "permissive" factors (McFarland, 1969a). In the case of pigeon reproductive behaviour, discussed by Vowles (1970), displacement preening occurs as part of the causal structure, although current theories of displacement activity would suggest that such behaviour is disinhibited (Rowell, 1961).

An alternative to correlation of sequences is temporal correlation, in which correlates between activities occurring within chosen intervals are calculated. This method suffers from the disadvantages that the result obtained varies with the time-interval chosen and that sequential information within the time-interval is lost (Blurton-Jones, 1968). A further disadvantage with both sequential and temporal correlations is that no account is taken of the fact that behaviour can vary in intensity as well as in duration or frequency. This aspect is well illustrated by reference to work on meal patterns.

A number of workers have found that the intervals which separate meals taken by rats, when food and water are available *ad libitum,* are positively correlated with the size of the meal which precedes them, but uncorrelated with the size of the meals which follow (Le Magnen and Tallon, 1966; Thomas and Mayer, 1968). The same appears to be true for chickens (Duncan *et al.,* 1970) and for doves (McFarland, unpublished). The evidence suggests that food intake is regulated by control of the length of intervals between meals, rather than by control of meal size. Thomas and Mayer (1968) obtained further evidence for this view by continuously infusing a small amount of liquid food directly into the stomach. They found that rats compensated by increasing the intervals between meals, without altering the meal size.

The finding that meal size is correlated with the interval following, rather than preceding, the meal, poses problems concerning the regulation of feeding behaviour. More especially, it is not clear how meal size is determined. Frequency histograms of meal sizes generally produce Poisson, or exponential, distributions, implying that successive meals are independent. For example, Duncan *et al.* (1970) obtained exponential distributions of meal sizes in chickens, the implication being that when a bird has started a meal the probability of stopping at any point is not markedly influenced by the amount already consumed. Thus chickens do not have a characteristic meal size. These workers also found that inter-meal intervals were exponentially distributed (see Fig. 5.29), indicating that such meals occur at random intervals.

On the evidence available, it is difficult to see how accurate regulation of energy balance is achieved. Part of the problem is that workers in this field have not used correlation functions which take into account both the temporal and intensity (meal size) aspects of the behaviour. As regulation must ultimately be geared to the nutrient properties of the diet, it is likely that these two aspects act jointly in determining the meal pattern. In other words, the time of occurrence and the size of a meal can be expected to be determined by the past history of meal taking.

A function which describes the general dependence of the values of the data at one time on values at another time is the "autocorrelation function". The autocorrelation function of a signal $x(t)$ is written $\phi_{xx}(\tau)$, and is defined as the time average of the product $x(t) \cdot x(t+\tau)$. That is, the product of the value at each point on the time-scale T, and a time τ units later (Fig. 5.33), is determined for each value of τ, and the average taken, viz.

$$\phi_{xx}(\tau) = \lim_{T \to \infty} \frac{1}{T} \int_{0}^{T} x(t) \cdot x(t+\tau)dt \qquad (5.32)$$

When $\tau = 0$, the time-history is correlated with an "image" of itself, and $\phi_{xx} = \psi_x^2$, the maximum value. As τ is increased the images may be seen as being displaced with respect to each other, and the value of $\phi_{xx} < 1$, but any periodicity in the data will tend to magnify ϕ_{xx}

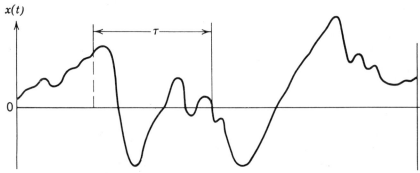

Fig. 5.33. Autocorrelation measurement. (From Bendat and Piersol, 1966.)

at the values of τ corresponding to the period. That for random data will have a high value only when $\tau = 0$ (Fig. 5.34).

In a preliminary study of meal patterns in doves, McFarland obtained autocorrelations for 12 h periods of *ad libitum* feeding and drinking, under controlled conditions of temperature (20°C) and daylength (12 h). The results, illustrated in Fig. 5.35, indicate that food and water intake are highly periodic, even though the probability density functions of meal intervals and sizes taken separately indicate that these variables are independent of their previous time-history, as found by other workers (Le Magnen and Tallon, 1966; Duncan *et al.,* 1970). The meal pattern thus appears to be dependent upon a joint function of temporal and intensity factors.

Figure 5.35 illustrates the autocorrelation functions obtained for both feeding and drinking. These are strikingly similar and this finding is in agreement with the view that *ad libitum* water intake is largely determined by food intake (Fitzsimons, 1958; Fitzsimons and Le Magnen, 1969; Kissileff, 1969; McFarland, 1969a). In general, the dependence of one set of data upon another can be determined by means of the crosscorrelation function, which is essentially the same as the autocorrelation function, except that the correlation is made between $x(t)$ and $y(t)$, viz.

$$\phi_{xy}(\tau) = \lim_{T \to \infty} \frac{1}{T} \int_0^T x(t) \cdot y(t+\tau)dt. \tag{5.33}$$

The crosscorrelation between feeding and drinking is illustrated in the lower part of Fig. 5.35. Note that the maximum value of ϕ_{wf} occurs at $\tau = 0$, which indicates that the effect of feeding upon drinking occurs within one time-unit, as might be expected from the fact that feeding and drinking are highly synchronous in this species (McFarland, 1969a). If the effect of feeding upon drinking were delayed, then the peck correlation would occur at a value of τ corresponding to the duration of the delay. Thus, while the autocorrelation function is particularly useful in determining periodicity hidden in random data, the crosscorrelation provides valuable information concerning the time-relations of different sets of data.

The power (mean square value) of a signal may be expressed in terms of the amplitude probability distribution (eqn. 5.26). In a noisy signal there may be various frequencies represented, and the variation of power with frequency is a measure of the "power

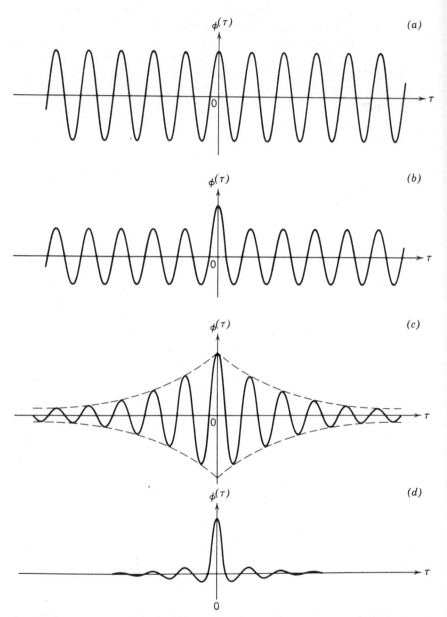

Fig. 5.34. Autocorrelation function plots: (a) Sine wave, (b) sine wave plus random noise, (c) narrow-band random noise, (d) wide-band random noise. (From Bendat and Piersol, 1966.)

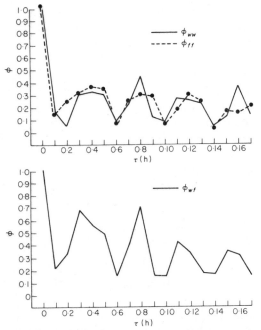

Fig. 5.35. Autocorrelation function plots obtained from records of *ad libitum* feeding (ϕ_{ff}) and drinking (ϕ_{ww}) in the Barbary dove. The crosscorrelation function plot between feeding and drinking (ϕ_{wf}) is indicated below. (From McFarland, unpublished observations.)

spectrum" of the signal. For mathematical convenience, the power is considered to be distributed equally between positive and negative frequencies and the "power spectral density" $\Phi(\omega)$ is chosen to satisfy the relationship

$$\sigma^2 = \frac{1}{2\pi} \int_{-\infty}^{+\infty} \Phi(\omega) d\omega. \tag{5.34}$$

The power spectrum of a signal contains the same information as the autocorrelation function and each may be obtained from the other by means of the Wiener-Khintchine equations

$$\phi(\tau) = \frac{1}{2\pi} \int_{0}^{\infty} \Phi(\omega) \cos \omega\tau \, . \, d\omega \tag{5.35}$$

$$\Phi(\omega) = 2 \int_{-\infty}^{+\infty} \phi(\tau) \cos \omega\tau \, . \, d\tau. \tag{5.36}$$

These two functions are "Fourier transforms" of each other. The Fourier transform, being similar to the Laplace transform, relates functions of time to functions of frequency (Bendat and Piersol, 1966; Douce, 1963).

One typical autocorrelation and power spectral density pair is of special interest. Consider a random signal in which every frequency is equally represented. This is called "white noise", and has a characteristic horizontal power spectrum. No correlation exists between different points on a white noise waveform because each value is, by definition, completely independent of all other values. Thus $\phi_{xx(\tau \neq 0)} = 0$, but when $\tau = 0$,

$$\phi_{xx}(\tau) = \lim_{T \to \infty} \frac{1}{T} \int_0^{+T} x^2(t)dt = \Psi^2 \delta(\tau). \tag{5.37}$$

Thus the autocorrelation function of white noise is an impulse of magnitude Ψ^2, and zero for $\tau \neq 0$. (See section 6.2.2).

In addition to empirical analysis of variable data, auto- and crosscorrelation functions are important theoretically in that they can be transformed from functions of time into functions of frequency, giving "power spectral density functions" and "cross spectral density functions", respectively (Bendat and Piersol, 1966). These functions, together with correlation functions, play an important role in the stochastic methods of system identification mentioned in section 3.2. Such stochastic methods are theoretically related to frequency analysis and they make possible an experimental approach that is particularly suitable for behavioural systems analysis, and is discussed further in Chapter 6.

5.2.4 Noise in the Pupillary and Accommodation Control Systems

The pupil exhibits random fluctuations in area, as well as low frequency drift. This aspect of the behaviour of the pupillary control system has been investigated by Stark (1968) and his co-workers, and provides some good examples of the application of some of the statistical methods discussed in this chapter.

During preliminary work, it was noticed that the pupil noise level varied with the stimulating light level, as indicated in Fig. 5.36. Experiments were carried out using the open-loop preparation discussed in section 4.2.4. Records of pupil area were obtained from the pupillometer and fed into a digital computer for analysis. Probability density functions for pupil area were obtained over a

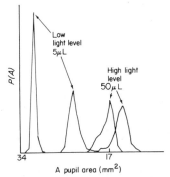

Fig. 5.36. Calibrated amplitude histogram of pupil area noise for low and high levels of light in microlumens. (From Stark, 1968.)

range of stimulation light intensity. These were found to be Gaussian in nature (Fig. 5.37) and the rms value of the pupil noise area signal (σ_A), and that of the derivative of the area ($\sigma_{\dot{A}}$), were found to be linear functions of average pupil area ($\overline{A}$), as shown in Fig. 5.38. To determine whether the illumination level or the average pupil area was the determining factor in the generation of the noise level, experiments were performed in which the light level was kept as low as possible and the pupil size changes by employing the accommodation response. When the focal plane of the fixation point is brought closer to the eye the average pupil area decreases, and it was found that rms area varied with pupil area under these conditions. This experiment demonstrates that the noise is introduced, or generated, in that part of the brain common to the pupil light reflex and the accommodation reflex.

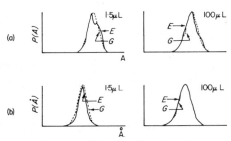

Fig. 5.37. Normalized amplitude histograms at various light levels in microlumens, for pupil area (A) and derivative of area (A). Solid lines indicate empirical results (E), dotted lines indicate fitted Gaussian curves (G). (From Stark, 1968.)

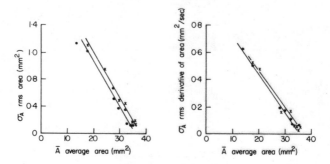

Fig. 5.38. Noise and noise derivative as a function of average pupil area showing linear relationships. Dots and crosses represent separate experiments. (From Stark, 1968.)

The linear relationship between pupil area and the rms area value suggests that the noise effects the system in a multiplicative fashion, and Stark (1968) proposed a simple model to account for this phenomenon (Fig. 5.39). It was also found that the autocorrelation function of the pupil noise was essentially the same for experiments

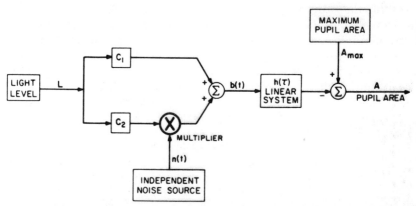

Fig. 5.39. Simplified model for pupil noise showing multiplicative gating of the noise source as well as a linear system to account for spectral characteristics of noise. (From Stark, 1968.)

conducted under high and low levels of illumination (Fig. 5.40). Near and far accommodation were found to make little difference to these functions. Crosscorrelation of the pupil noise from the left and right eyes indicates that the noise is introduced or generated at a point in the brain common to both eyes. Combined with the previous

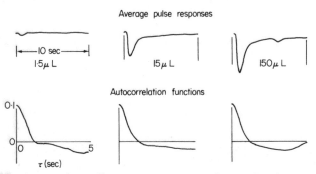

Fig. 5.40. Averaged pupil area responses and associated autocorrelation function plots; light pulse amplitudes in microlumens above a constant background. (From Stark, 1968.)

evidence, this suggests that the Edinger-Westphal nucleus is the main source of noise in this system.

From these and other experiments Stark (1968) is able to modify the model based upon transient and frequency analysis (Fig. 4.14), so as to incorporate the characteristics of the system that are revealed by statistical analysis of the random fluctuations in pupil area. This model is illustrated in Fig. 5.41. Thus, statistical methods can serve both as a means of removing "unwanted" variation and as an experimental tool in systems analysis.

Campbell *et al.* (1959) measured fluctuations in accommodation

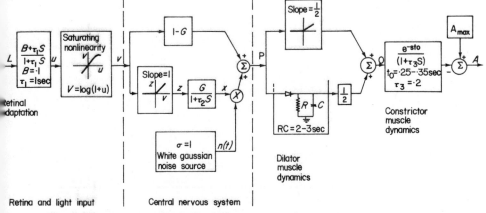

Fig. 5.41. Model of pupil reflex system accounting for pupil noise phenomena. Accommodation input could be injected at V. Compare with Fig. 4.14. (From Stark, 1968.)

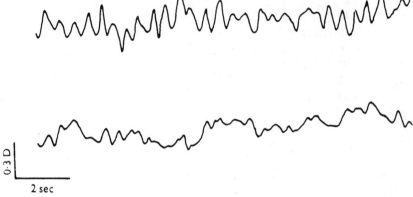

Fig. 5.42. Accommodation record of one subject under normal viewing conditions with a 7 mm pupil (upper) and with a 1 mm effective entrance pupil of the eye (lower). The records have the same average accommodation level. (From Campbell *et al.*, 1959.)

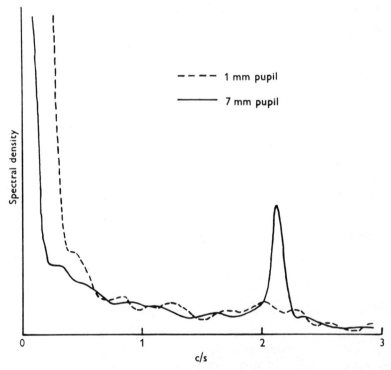

Fig. 5.43. Frequency spectra of the two records shown in Fig. 5.42. (From Campbell *et al.*, 1959.)

under steady viewing conditions, using an iris diaphragm as the effective entrance pupil of the eye. Records of the fluctuations in accommodation were fed into a digital computer and their autocorrelation functions obtained. These were subjected to a Fourier transformation to obtain the spectral density functions. Sample records of the accommodation fluctuations are illustrated in Fig. 5.42, and their frequency spectra in Fig. 5.43. In the wide pupil experiment there is a well marked high frequency component which

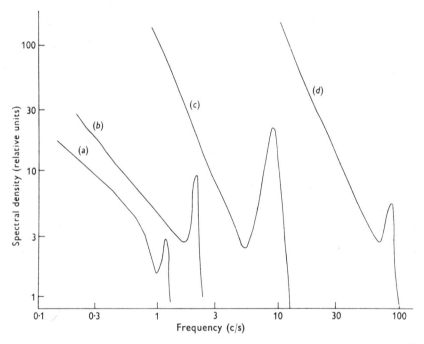

Fig. 5.44. Frequency spectra of various motor systems in a steady environment, drawn to show principle features: (a) pupil diameter during steady illumination of the retina, (b) accommodation during steady viewing of a near target, (c) finger displacement during steady pointing, (d) eyeball position during steady fixation. (From Campbell *et al.*, 1959.)

is greatly diminished in the narrow pupil experiment. Similar high frequency components have been demonstrated in records of pupil diameter, finger movement and eyeball position (Fig. 5.44), though their significance is not fully understood.

Fluctuations in accommodation have been suggested as the cause of the moving visual images produced by some regular stationary

Fig. 5.45. Concentric ring target of Helmholtz. Rotating sectors can be seen when the target is viewed with one eye, from a distance of about 1 ft. The time course of the movements corresponds to that of fluctuations of accommodation shown in Fig. 5.42. (From Campbell *et al.,* 1959.)

patterns. For example, when the concentric circles of Fig. 5.45 are viewed with one eye, from a distance of about 1 ft, rotating sectors can be seen. Helmholtz in 1856 (Helmholtz, 1924) attributed this to changes in accommodation and some more recent authors (e.g. Campbell and Robson, 1958) are of the same opinion.

CHAPTER 6

Black Box Analysis

Most psychologists are familiar with the representation of an animal as a "black box". This conception arose partly because of the influence of cybernetics and partly from the "empty organism" approach (Boring, 1946) developed by many academic psychologists in their revolt against introspective psychology. The extreme position in this respect was reached by Skinner (1938, 1950) in his disapprobation of all theorizing that is not directly related to the functional relation between the manipulation of stimuli and the resulting observed responses. But now, most students recognize the value of employing intervening variables as aids to the simplification of description and explanation.

The aim of this chapter is not to argue the status of "intervening variables" and "hypothetical constructs" (MacCorquodale and Meehl, 1948), or to question the utility of block diagrams versus the "real nervous system" (Weiskrantz, (1968). Such matters do raise questions that are important for behaviour theory and some of these are discussed in the preface to this book. The present aim is to discuss some of the problems that arise in black box analysis in general, to relate these to behavioural studies where possible, and to point to some of the problems that are likely to arise in behavioural systems analysis in the future.

The strategies employed in the investigation of biological systems from the viewpoint of control theory are of two main types: the analytic and the synthetic. The analytic approach primarily consists in the investigation of the input-output relationships of the system. Inputs of known waveform are applied and correlated with the measured outputs. For simple systems transient analysis is adequate, and examples of this approach are discussed in previous sections (sections 3.2.3 and 4.2.3). For higher-order systems frequency analysis is more suitable, and Chapter 4 is devoted to this topic. Stark and Young (1964) have pointed out some of the dangers of applying such methods indiscriminately to biological systems. In particular, slow-adaptation (Stark, 1959), precognitive input prediction (Stark et al., 1962) and task adaptation (Young et al., 1963), have been demonstrated in quasi-behavioural situations. Although the presence of such phenomena can sometimes invalidate certain methods of investigation, a large battery of systems analysis techniques is now available and alternative methods of analysis can usually be found. The tactical aspects of systems analysis are largely concerned with the employment of techniques designed to avoid, or overcome, the non-linearities in the system.

The synthetic approach consists in putting together elements or sub-systems of known input-output characteristics. The functional relationships between elements prescribe the behaviour of an hypothetical or abstract system, the validity of which can be tested experimentally. This approach is of great value in integrating the interactions between the components of large and complex systems, where the components have previously been studied in isolation. Recent applications of such methods to systems controlling body temperature (Crosbie et al., 1961), respiration (Milhorn, 1966) and cardiovascular mechanisms (Grodins, 1959), herald a new integrative approach to the physiology of homeostatic mechanisms. Synthetic methods generally involve computer simulation of the system under study and offer what is, at present, the only quantitative method of dealing with unavoidable non-linearities.

When the behaviour of a system can be described by a set of linear differential equations, the system can be assumed to be linear for practical purposes (see section 1.3). In reality however all systems are to some extent non-linear. A linear system should theoretically produce its normal type of response to an infinitely small input but in practice this is never so. Similarly, there is always a limit to the magnitude of the input that will produce a linear response. For example, Chung (1965) showed that the response rate of pigeons pecking in a Skinner box changes when the force required to operate the key is altered and then returns to the normal level. This is true only for force requirements up to a critical value above which the animal is unable to maintain its normal steady-state level of responding (Fig. 6.1). McFarland (1966c) showed that this behaviour could be accounted for in terms of the first-order feedback system illustrated in Fig. 6.2. The system is linear, except for the saturation on the controller, which prevents the applied force from rising above a certain value. As long as the force requirement remains below the critical value the system behaves in a linear fashion, producing

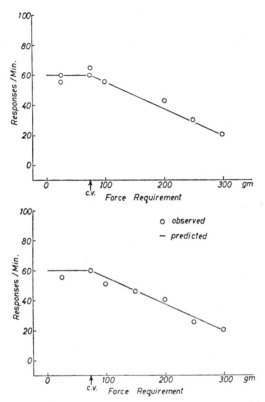

Fig. 6.1. Steady-state operant response rate as a function of force required to operate the key. *c.v.* = critical value. Data from Chung (1965). (From McFarland, 1966c.)

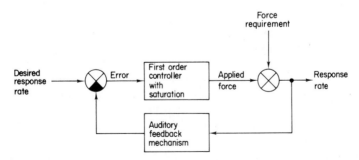

Fig. 6.2. Outline of McFarland's model for control of response rate in a Skinner box.

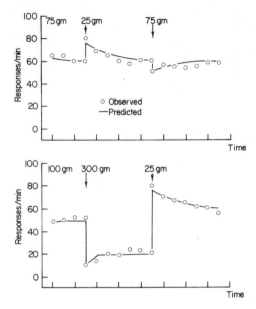

Fig. 6.3. Changes in response rate which result from changes in force requirement (shown by numbered arrows). (a) Shows data in the linear (non-limited) region, (b) behaviour in the non-linear (limited) region. (From McFarland, 1966c.)

first-order transient responses following changes in the force requirement (Fig. 6.3a). Above the critical value, the system behaves in the non-linear fashion illustrated in Fig. 6.3b.

Non-linearities, such as this, are extremely common in biological systems and their existence provides considerable problems for the analysis of such systems as there are no generally applicable mathematical methods by means of which non-linear systems can be described or analysed. Three main strategies exist by which these problems can be tackled. (1) The non-linearities can be avoided, by small-signal analysis (section 4.2.4), or by "linearizing" the behaviour of the system by careful choice of the time-unit used in analysis (section 3.1.3). (2) The non-linearities can be tackled analytically by the use of graphical or equivalent methods, instead of conventional mathematics (section 6.1.3). (3) The non-linearities can be incorporated into a computer simulation of the system, the aim being to synthesize a model which behaves like the system under investigation in every respect. The first of these approaches has been discussed in previous chapters (sections 3.1.3, 3.3, 4.2.4). The other two divide into analytic and synthetic strategies, within which various tactics may be employed.

Before proceeding to empirical methods of black box analysis there are a number of questions concerning the logical status of the method, which need to be considered. Specifically, what types of hypothesis can be validated by black box analysis, and what kind of explanation is offered by such an exercise?

The former problem received early attention from Ashby (1956), has since been developed by control engineers (Kalman, 1960; Kreindler and Sarachik, 1964; Schultz and Melsa, 1967), and is discussed in the next section. The second question, in the context of behavioural systems analysis, is essentially a question concerning psychological theory, and is dealt with in the next chapter.

6.1 MATHEMATICAL REORIENTATION

The classical approach to systems analysis involves the use of transfer functions to describe input-output relationships. This approach has been used in this book for three main reasons: it is well suited to the analysis of linear systems, is relatively easy to understand, and is almost the only approach that has been used by biologists. Biological systems, however, have a number of characteristics for which the transfer function approach is not well suited: they tend to be multi-input-output systems, which are non-linear and noisy. Modern control theory embraces many methods, developed in response to the challenge offered by systems with these properties. While it is beyond the scope of this book to describe these methods in any detail, it is hopefully worth while at least to touch upon the type of thinking that will come to dominate biological systems analysis in the future. And the methods outlined here have immediate practical relevance in that they offer solutions to some of the problems that the experimenter faces in the laboratory.

6.1.1 State Variables

A linear system can be described in terms of a transfer function in which a number of state variables can be identified. By definition, a transfer function is the Laplace transform of a differential equation with zero initial conditions. The order of the differential equation is the same as the number of state variables and the number of initial conditions that must be specified in order to predict the future behaviour of the system (section 3.2.1). Representation of linear systems in state variable form involves n first-order differential equations, as opposed to the usual nth order equation (Schultz and Melsa, 1967). Thus the first step in deriving a state variable representation from a transfer function is to obtain the n first-order differential equations. For example, the transfer function

$$H(s) = \frac{y(s)}{x(s)} = \frac{K}{s^3 + k_1 s^2 + k_2 s + k_3} \tag{6.1}$$

is equivalent to the following third-order differential equation

$$\frac{d^3 y}{dt^3} + k_1 \frac{d^2 y}{dt^2} + k_2 \frac{dy}{dt} + k_3 y = Kx(t) \tag{6.2}$$

Let $y(t) = a, \qquad \dfrac{dy}{dt} = b, \qquad \dfrac{d^2 y}{dt^2} = c$ \hfill (6.3)

then

$$\frac{dc}{dt} + k_1 c + k_2 b + k_3 a = Kx \tag{6.4}$$

giving the three first-order differential equations,

$$\frac{da}{dt} = b \tag{6.5}$$

$$\frac{db}{dt} = c \tag{6.6}$$

$$\frac{dc}{dt} = Kx - k_1 c - k_2 b - k_3 a \tag{6.7}$$

The variables b and c, being derivatives of the system output a, are known as "phase variables". In general, phase variables are defined as those state variables which are obtained from one of the system variables and its $n-1$ derivatives (Schultz and Melsa, 1967). The most commonly used variables are the system output and its derivatives. The phase variable representation of a system can be portrayed in block diagram form, as illustrated in Fig. 6.4. Phase variables do not generally correspond to physical variables that can be measured or manipulated experimentally. Although phase variables are useful in certain types of analysis, particularly in describing the behaviour of certain types of non-linear systems in terms of the phase plane (section 6.1.3), they do not enable the

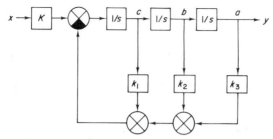

Fig. 6.4. Phase variable representation of a third-order system.

investigator to take equal account of all the important state variables in a system. A more intuitive approach is to consider those system variables that are physically meaningful.

A system transfer function can generally be broken up into a number of sub-systems each containing a single integrator. There will generally be a number of different ways in which this can be done, as illustrated in Fig. 6.5. It is not possible to say whether the state variables z_1 and z_2 are physically meaningful, without further knowledge about the system. If the complete block diagram of the system is known then it is a simple matter to choose physically meaningful state variables. If a black box problem is involved then identification of physically meaningful variables is less easy.

A good starting point is to identify the input and output variables in terms of generalized flow and effort variables (see section 1.1). In Fig. 6.5a, the input variable x may be known to be voltage, an effort

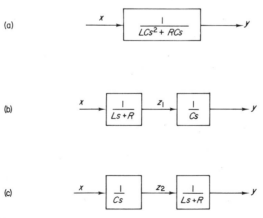

Fig. 6.5. Alternative ways of dividing a second-order system into two first-order sub-systems.

variable. This, applied to an integrator gives a flow variable and the integrator must take the form of a generalized inductance. For example, a voltage applied to an inductance produces a current. Therefore, the first sub-system of Figs 6.5b or c must contain an inductance, and either z_1 or z_2 must correspondingly represent the flow variable. If from the behaviour of the system it can be deduced that the resistance must be associated with the inductance, so that Fig. 6.5b represents the correct formulation, then z_1 is physically meaningful, whereas z_2 is not.

Having identified the relevant effort and flow rate variables, the corresponding state variables can be obtained directly. Thus generalized displacement is the integral of the flow variable and generalized momentum is the integral of the effort variable. In terms of the electrical system of Fig. 6.5 these generalized state variables correspond to the charge and the flux-linkage, respectively (see section 1.1). Generalized state variables can be identified by analogy (Olson, 1958), or from first principles (Macfarlane, 1964), and a summary of their interrelations is given in Fig. 6.6.

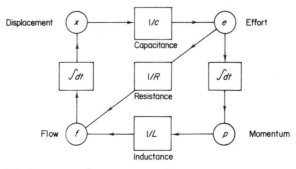

Fig. 6.6. Summary of interrelations between generalized variables.

The generalized variables of physical systems can be carried over into biological systems analysis (Milsum, 1966), and the terminology has been carried over into the description of motivational systems (McFarland, 1970c), as is described in the next chapter.

6.1.2 Controllability and Observability

Although state variables can be identified from the system transfer function, it is possible, particularly in multi-input-output systems that important state variables could be overlooked. This possibility raises the question of the general relationship between the transfer-function and state variable representations of a system. It is especially important to know whether a mathematical description (realization) of a system is "minimal" in the sense that the system is realized in the simplest possible manner, with the minimal number of parameters, etc. Kalman (1963, theorem 7) has shown that realizations are minimal if, and only if, they are completely controllable and completely observable. Complete "controllability" means that it is possible to move any state to any other state during any finite interval of time by a suitably chosen input. Complete "observability" means that knowledge of the input and output over any finite interval is sufficient to determine the state of the system uniquely (Kalman, 1968).

Uncontrollable states are those that are completely decoupled from the system input, while unobservable states are decoupled from the output. Gilbert (1963) has shown that systems may be divided into sets of sub-systems having different controllability and observability properties. Thus a sub-system may be (1) controllable and observable, (2) controllable but unobservable, (3) uncontrollable but observable, (4) uncontrollable and unobservable. These distinctions are illustrated in Fig. 6.7. From this figure it can be seen that the transfer function relates only to those parts of a system that are both controllable and observable and it thus provides a minimal realization of such sub-systems. Similarly, the phase variable representation is always controllable and observable, and it is possible to show (Kalman, 1963) that any controllable and observable system may always be represented in phase variable form.

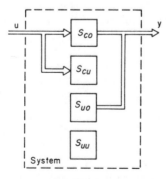

Fig. 6.7. Partitioning a system on the basis of controllability and observability criteria. Thick arrows indicate matrix relationship (see Fig. 6.10). (From Schultz and Melsa, 1967.)

The concepts of controllability and observability have implications for black box analysis that are only just beginning to be realized. In particular, the concepts are important in assessing the status of models (Kalman, 1968), and in the application of an approach, known as Kalman filter theory (Schultz and Melsa, 1967), to the estimation of inaccessible state variables.

In behavioural systems it is unlikely that any sub-system will be completely controllable and observable, and the transfer function can therefore only be regarded as an approximation. Uncontrollable sub-systems, such as those responsible for endogenous rhythms, affect many aspects of behaviour (Aschoff, 1963), and unobservable sub-systems are important in any process in which information is stored for future reference, without any change in behaviour being involved.

6.1.3 State Representation

The state of a system is defined by the state variables which together with the system inputs are sufficient to determine the behaviour of the system (see section 3.2.1). The generalized state variables, displacement and momentum, are both "vector" quantities, having both magnitude and direction. The state of a system can thus be represented in a vector space, the archetypal form of which is linear and has cartesian coordinates. Thus the state of a second-order system could be represented as a point in two-dimensional space, the coordinates of which correspond to the two state variables. In the case of the second-order system discussed in section 3.3.2 (Fig. 3.21), the relevant state variables are water debt and gut content. In practice, the state of a system changes continually with time, and the point representing this state traces a "trajectory" in the state space. In the present example, the transient behaviour of the state variables can be determined from the block diagram (section 3.3.2) and portrayed in the two-dimensional state space, as shown in Fig. 6.8.

In the case of phase variables, it is common practice to plot the system output and its derivative in a two-dimensional space, called the "phase plane". This approach is particularly useful in the analysis of second-order non-linear systems. As an example, consider the typical second-order transient illustrated in Fig. 3.14. The figure shows the position of the wrist, as a function of time, following an impulse delivered in a simple pointing task (Fig. 3.13). In the phase

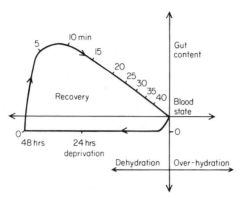

Fig. 6.8. State plane trajectory for Fig. 3.21 step response. (From McFarland, 1970c.)

plane the wrist position is plotted against its derivative, velocity, as shown in Fig. 6.9b. By convention, the trajectory moves clockwise, and the final steady-state is taken as the point at which the axes cross. The wrist movement is typical of a second-order underdamped system (see section 4.1.4), and can be represented in the following form

$$\ddot{x} + a\dot{x} + bx = 0 \qquad (6.8)$$

where a and b are constants, x is the dependent variable, $\dot{x}$ and $\ddot{x}$ are the first and second derivatives respectively. Thus

$$\frac{dx}{dt} = \dot{x}, \qquad (6.9)$$

and from eqn. (6.8)

$$\frac{d\dot{x}}{dt} = -a\dot{x} - bx. \qquad (6.10)$$

Dividing eqn. (6.10) by eqn. (6.9) gives

$$\frac{d\dot{x}}{dx} = -a - \frac{bx}{\dot{x}}. \qquad (6.11)$$

The trajectory on the phase plane can be constructed by finding those lines, called "isoclines", along which the slope of the trajectories is constant. Thus, from eqn. (6.11)

$$k = -a - \frac{bx}{\dot{x}} \qquad (6.12)$$

which is the equation for the isocline k, and

$$\frac{\dot{x}}{x} = \frac{-b}{k+a}.$$
(6.13)

From eqn. (6.13) the isoclines specified by k can be constructed on the phase plane, as a series of lines of slope $-b/(k+a)$. For instance, when $k = 0$ the isocline is the line of slope $-b/a$ passing through the origin, as illustrated in Fig. 6.9b. This isocline, at which the

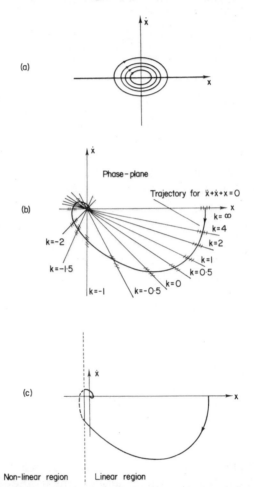

Fig. 6.9. Representation of second-order transient responses in the phase plane. (a) Undamped system with various amplitudes represented. (b) Underdamped system response, with isoclines indicated. (From Sensicle, 1968.) (c) Underdamped system with a hard (solid line), or a soft (dashed line) limiter.

trajectory is parallel to the x axis, is often called the "change-over boundary". Its slope increases as the damping is reduced, and in a completely undamped system the change-over boundary is coincident with the x axis, so that the trajectory forms an elipse (Fig. 6.9a).

The trajectory of a linear underdamped second-order system is illustrated in Fig. 6.9b. This trajectory is typical of a damped oscillatory wrist movement eventually coming to rest. Suppose that the movement of the wrist were constrained by a metal stop beyond which the pointer could not move, so that angular positions above a certain value were impossible. The output of the system would then be limited and the impulse response would be non-linear, something like that illustrated in Fig. 6.9c. In the phase plane, the trajectory takes its normal course as long as it remains within the linear region (Fig. 6.9c). In the present case the non-linear region is not entered, the output of the system being contstrained by a "hard limiter". If the pointer were constrained by a soft rubber cushion, rather than a metal stop, the output would be subject to a "soft limiter", and the non-linear region of the phase plane would be entered (Fig. 6.9c). This form of non-linearity is particularly common in biological systems. The stomach, for example, acts as a soft limiter when its contents reach a certain volume.

Phase plane methods are useful in the analysis of many other types of non-linearity, such as threshold and switching phenomena. Accounts of this approach are given in books on control theory (e.g. Macmillan, 1962; Sensicle, 1968), in far greater detail than is possible here. As well as its purely practical value, the phase plane is a particular case of the more general state-space approach. So far only two-dimensional spaces have been considered but state-space methods are applicable to systems having any number of state variables, the state of such systems being represented in an n-dimensional space.

Consider a system having two inputs and two outputs, as shown in Fig. 6.10. If the system is linear, the relationship between the inputs and outputs can be represented by means of two equations

$$y_1 = x_1 k_1 + x_2 k_2 \tag{6.14}$$

$$y_2 = x_1 k_3 + x_2 k_4, \tag{6.15}$$

where k_1, k_2, k_3, k_4 are transmittances relating to each input-output possibility. These two linear equations can be written as a single "matrix equation"

$$\begin{bmatrix} x_1 \\ x_2 \end{bmatrix} \begin{bmatrix} k_1 & k_2 \\ k_3 & k_4 \end{bmatrix} = \begin{bmatrix} y_1 \\ y_2 \end{bmatrix}. \tag{6.16}$$

The transmittances k_1, k_2, k_3, k_4 are the "elements" of the matrix. They are ordered detached coefficients, which have an algebra of their own, called "matrix algebra". The inputs and outputs are represented by pairs of ordered numbers, called "column matrices".

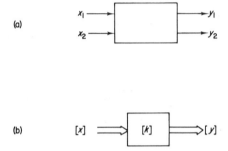

Fig. 6.10. A dual input-output system. (a) Normal representation, (b) matrix representation.

These can be represented in a vector space, just as state variables can. A pair of inputs can be represented on a plane, as in Fig. 6.11 and n inputs could similarly be represented in an n dimensional vector space. Figure 6.11 shows that the inputs, taken as a whole, can be represented as a vector having magnitude and direction dependent upon the values of the input variables at each point in time. Thus the input and output column matrices are often called "column vectors".

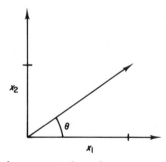

Fig. 6.11. Vectorial representation of a two-member column matrix.

The elements of the matrix eqn. (6.16) form a 2 × 2 matrix, which can be denoted by the single letter [K]. Similarly, the input and output column vectors can be represented by [X] and [Y] respectively. Equation (6.16) can thus be expressed in the following form

$$[Y] = [K] [X] \qquad (6.17)$$

The multiplication of the matrix [K] by the vector [X] is non-commutative and must be carried out in accordance with the laws of matrix algebra. It is important, therefore, to distinguish symbols representing matrices from those representing ordinary numbers. This is often done by printing the former in bold type, or by enclosing them in squared brackets. For information concerning the rules of matrix algebra, the reader is referred to an elementary text (e.g. Coulson, 1969). Summaries are given in some books on control theory (e.g. Brown, 1965; Elgerd, 1967; Schultz and Melsa, 1967).

Representation of an nth-order system in matrix form can be achieved by converting the nth-order differential equation into a set of n first-order differential equations, as illustrated in section 6.1.1. A general form of such a set of equations is

$$\begin{aligned}
x'_a &= a_1 x_a + a_2 x_b + \ldots a_n x_r + k_a \\
x'_b &= b_1 x_a + b_2 x_b + \ldots b_n x_r + k_b \\
&\quad \ldots \ldots \\
x'_r &= r_1 x_a + r_2 x_a + \ldots r_n x_r + k_r
\end{aligned} \qquad (6.18)$$

These equations may be expressed in matrix form as follows,

$$\frac{d}{dt}\begin{bmatrix} x_a \\ x_b \\ \\ x_r \end{bmatrix} = \begin{bmatrix} a_1 & a_2 \ldots a_n \\ b_1 & b_2 \ldots b_n \\ \ldots \ldots \\ r_1 & r_2 \ldots r_n \end{bmatrix}\begin{bmatrix} x_a \\ x_b \\ \\ x_r \end{bmatrix} + \begin{bmatrix} k_a \\ k_b \\ \\ k_r \end{bmatrix} \qquad (6.19)$$

alternatively, $\dfrac{d[\mathrm{x}]}{dt} = [K] [\mathrm{x}] + [\mathrm{k}]$ \qquad (6.20)

The matrix representation applies equally to functions of the Laplace operator (s), as to time functions. Thus a multi-input-output system could be represented in the following form

$$[\mathrm{y}](s) = [G](s)[\mathrm{x}](s) \qquad (6.21)$$

where $[G](s)$ is known as the "transfer function matrix". Thus, the element $g_{nr}(s)$ is the transfer function between the rth input and the nth output, viz.

$$g_{nr}(s) = y_n(s)/x_r(s). \tag{6.22}$$

Any linear system can be represented in matrix form, as can some types of non-linear system. The main value of the matrix method is that it is well suited to the representation of multi-input-output systems and to the description of systems in terms of state variables, rather than transfer functions. This is a general trend in modern control theory, and interested readers are referred to textbooks on this subject for further information (e.g. Brown, 1965; Schultz and Melsa, 1967). The main purpose of discussing these topics here is to illustrate the point that any system which can be described in terms of sets of numbers (e.g. a number of state variables, or of inputs and outputs), can be represented in an n-dimensional Euclidean space. In addition to providing a useful descriptive tool, this approach also makes possible the use of powerful analytical tools, as matrix algebra and vector analysis.

6.2 IDENTIFICATION

The analysis of control systems involves three distinct types of problem (Milsum, 1966):

(1) The "output problem": given the input and a description of the system, find the output. For linear systems this may be a straightforward matter of solving the relevant equations by direct calculation, or by flow graph simplification, as outlined in section 1.4, and exemplified in sections 2.3 and 3.3.1. For non-linear systems, similar methods can be used if the system can be suitably linearized. A more direct approach is the "slope-line method" (Milsum, 1966; Naslin, 1962, 1965) which is a graphical implementation of the numerical analysis used in digital computation.

(2) The "input problem": given a description of the system, and a sample output, find the corresponding input function. This problem can often be solved by inverting the transfer function and regarding the output as an input to the inverted system. Thus if

$$y(s) = H(s) \cdot x(s) \tag{6.23}$$

where $x(s)$ is the unknown input and $y(s)$ the known output, then

$$x(s) = H^{-1}(s) \cdot y(s). \tag{6.24}$$

By hand, this may be a tedious calculation, but by using an analogue computer (see section 6.3.1) it becomes a relatively trivial exercise (Milsum, 1966).

(3) The "identification problem": given samples of the input, and corresponding output, functions, find the transfer function. This is the central problem in systems analysis. For single-input-output linear systems the problem is relatively straightforward. The transfer function may be determined by transient analysis (section 3.2) or frequency analysis (section 4.1). The inverse Laplace transform of the transfer function gives the "impulse response" of the system. That is, the response that the system, having zero initial conditions, would produce when subjected to an ideal impulse input (see section 3.2.2). The impulse response is sometimes more convenient to obtain than the transfer function; either directly, by subjecting the system to a single short pulse input, or by correlation techniques as illustrated in section 6.2.2. The reflex wrist response illustrated in Fig. 3.14 is an example of an impulse response obtained by the former method.

For systems with multiple inputs and outputs, and for noisy systems, analytical identification procedures become more difficult; while for non-linear systems they are almost useless. Problems of the former type may be tackled by a combination of stochastic and matrix methods, while simulation is probably the best way to approach highly non-linear systems.

6.2.1 Single-input-output Systems

The complexity of biological systems necessitates a high degree of selection on the part of the investigator. Having isolated a sub-system that is suitable for study the biologist has two options. He can identify a suitable input variable and a suitable output variable and try to determine the relation between the two, while holding other input variables as constant as possible. Alternatively, he can identify a number of input and output variables and proceed by a method appropriate to the analysis of multiple-input-output systems. Most biological workers have employed the former procedure. Thus although all biological systems are, by virtue of their complexity, multiple-input-output systems, techniques appropriate to single-input-output systems have been most commonly employed. Consider, for instance, the two examples of frequency analysis discussed in Chapter 4. In the case of the determination, by McFarland and Budgell (1970b) of the transfer function relating ambient temperature and operant drinking behaviour in doves (section 4.1.2), it is well known that temperature is only one of the many factors affecting drinking. Factors such as water balance and the effect of food intake can be controlled by maintaining the animals on a regular deprivation schedule, under controlled environmental conditions. The effects of water ingestion during the test itself are allowed for by running the animals on a random interval reward schedule (VI 2 min), so arranged that changes in the response rate make no difference to the number of rewards obtained. Because the rewards arrive at random intervals, and all calculations are in terms of frequency, the satiation feedback loop is effectively opened by this procedure. In the case of Stark's (1968) frequency analysis of the pupillary light reflex (section 4.2.4), the feedback loop is also opened experimentally. Nevertheless, the pupil response exhibits high frequency noise and low frequency drifting (Fig. 4.18). Under such conditions it may be difficult to obtain reliable phase comparisons between input and output and special phase detecting procedures may have to be used. For example, cross-correlation of input and output will give an accurate indication of the phase lag (see section 5.2.3).

Cross-correlation of input and output forms the basis of a number of methods of system identification. Just as the auto-correlation function can be transformed mathematically into the power spectral density function (section 5.2.3), the cross-correlation function can be similarly transformed into the "cross spectral density" function (for details see Bendat and Piersol, 1966; Korn, 1966). A plot of the cross spectral density function versus frequency, called a "cross spectrum", has both gain and phase components, similar to those of a Bode plot. Thus cross spectral density functions can be used to provide frequency response functions. For a linear system with stationary input $x(t)$ and output $y(t)$

$$H(\omega) = \frac{\Phi_{xy}(\omega)}{\Phi_{xx}(\omega)},$$ (6.25)

where $H(\omega)$ is the frequency response function, $\Phi_{xy}(\omega)$ is the cross spectral density function, and $\Phi_{xx}(\omega)$ is the power spectral density function.

In the behavioural context, the definition of input and output may be to some extent arbitrary, and correlations between different behaviour patterns can be used to determine the nature of their mutual interaction. Thus Delius (1969) determined the power spectral density functions for flying and for comfort movements in the skylark (*Alauda arvensis*), observed under natural conditions. The power spectrum for comfort behaviour (Fig. 6.12a) indicates a high proportion of low frequency components and two series of equi-spaced maxima (6, 13, 20, 27 cyc/ and 3, 10, 17 and 24 cyc/h), suggesting that rhythmical processes play an important part in determining the appearance of the behaviour. The power spectrum

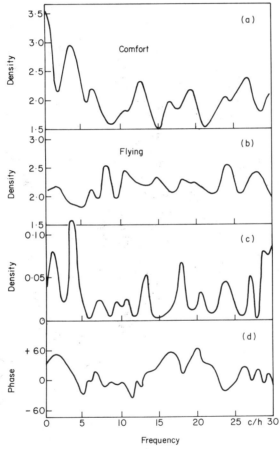

Fig. 6.12. Spectral and cospectral density functions for comfort behavior and flying on the skylark (*Alauda arvensis*). (From Delius, 1969.)

for flying (Fig. 6.12b) is more similar to that of white noise, covering a wide range of frequencies. The cross spectrum between flying and comfort behaviour (Fig. 6.12c and d) is diminished in magnitude, indicating that the sharing of activity between the two is restricted.

The low frequency peaks suggest that out-of-phase relationships predominate, while at high frequencies the preponderant relationships seem to be in-phase. In general, it seems that flying and comfort behaviour are to a large extent statistically independent processes, with minimum common causal factors, and little interaction. Nevertheless, Delius' study is important in that it points the way towards the usefulness of stochastic methods in the analysis of naturally occurring behaviour patterns.

The complexity of behaviour makes interpretation of correlation and spectral functions a difficult business, especially where no manipulative procedures are involved. Causal relations cannot be inferred from correlation functions alone. On the other hand, manipulation of behavioural variables may disrupt the normal functioning of the system. Thus transient inputs, when large enough to overcome the noise in the system, often drive variables beyond their normal limits and may induce artefactual non-linearities. Periodic inputs are suitable for some types of system, but many behavioural systems are able to anticipate periodic stimulus changes and the consequent predictive behaviour tends to mask the true nature of the system. Stark (1968) obtained quite different Bode plots for eyeball tracking of predictable and unpredictable targets (Fig. 6.13). Unpredictable inputs can be generated by superposition of sinusoidal functions and by random or pseudo-random functions. The use of random input functions, together with stochastic methods of analysis, has proved to be a powerful tool in system identification.

6.2.2 The Impulse Response

The response of a linear system to a unit impulse $\delta(t)$ is called the "impulse response" $h(t)$. Since the Laplace transform of the unit impulse is 1 (section 3.2.2), the transform of the response of a system to a unit impulse is identical to the transfer function $H(s)$ of the system. Thus where $y(t) = h \cdot \delta(t)$, $y(s) = H(s)$, and $h(t) \rightleftharpoons H(s)$. The response $y(t)$ of a system to an input function $x(t)$ can be given by the "convolution integral"

$$y(t) = \int_{-\infty}^{+\infty} h(t-u)x(u)\,du = \int_{-\infty}^{+\infty} h(u)x(t-u)\,du \qquad (6.26)$$

where $x(u)$ is the input at any time u. The cross-correlation between $x(t)$ and $y(t)$, given by eqn. 5.33, is

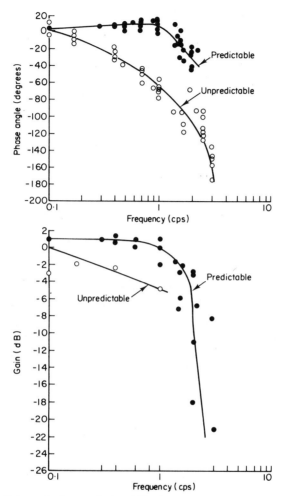

Fig. 6.13. Gain and phase relationships for continuous predictable (sinusoid) and unpredictable (sum of four sinusoids) target motions in eye-tracking experiments. (From Stark, 1968.)

$$\phi_{xy}(\tau) = \lim \frac{1}{T} \int\limits_0^T x(t) \cdot y(t+\tau)\,dt$$

$$= \lim \frac{1}{T} \int\limits_0^T x(t) \int\limits_{-\infty}^{+\infty} h(u)x(t-u+\tau)\,du\,dt. \qquad (6.27)$$

Interchanging the order of integration

$$\phi_{xy}(\tau) = \int_{-\infty}^{+\infty} h(u) \left\{ \lim \frac{1}{T} \int_{0}^{T} x(t) . x(t-u+\tau)dt \right\} du . \tag{6.28}$$

From eqn. 5.32

$$\phi_{xy}(\tau) = \int_{-\infty}^{+\infty} h(u) \, \phi_{xx}(\tau-u)du \tag{6.29}$$

provided that u and τ are approximately equal.

Thus cross-correlation between input and output yields the impulse response of the system, multiplied by the auto-correlation function of the input. This equation, often called the Wiener-Lee relation, is fundamental to the stochastic method of identification. Consider the case when the input is a white noise signal. The auto-correlation of white noise is an impulse function (eqn. 5.37); and the cross-correlation function, between this input and the output of the system, will therefore be proportional to the impulse response of the system. This method of identification has been used in a few biological studies (e.g. Stark, 1968), but usually the white noise function is not convenient for biological experimentation as it is difficult to produce. The difficulty can be overcome by the use of a pseudo-random binary sequence (PRBS), such as that illustrated in Fig. 6.14a. This function has a triangular auto-correlation function

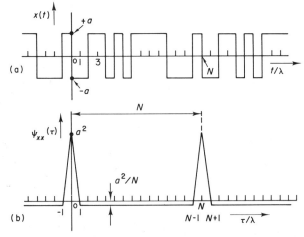

Fig. 6.14. (a) Pseudo-random binary sequence. (b) Auto-correlation function plot of (a). (From Hughes, 1969.)

(Fig. 6.14b) which can be made to approximate an impulse by suitable choice of input parameters (Korn, 1966).

The PRBS method has a number of advantages which make it particularly well suited to behavioural systems analysis: (1) The fact that it is a binary function means that it can be produced easily, compared with a sinusoidal function for example. The form "stimulus on–stimulus off" is more likely to be relevant to the behavioural situation than a continuously varying function. (2) Determination of the impulse response can be carried out in a single experimental session, whereas frequency analysis normally requires a separate session for each frequency employed. (3) The PRBS can be small in amplitude, thus inducing minimal disturbance during normal operation of the system. (4) The repetitive nature of the test signal makes the correlation function particularly immune to noise in the system. The main disadvantages are that non-linearities in the system are difficult to spot and may lead to misinterpretation of the results and, secondly, that a considerable amount of calculation is required and access to a digital computer necessary.

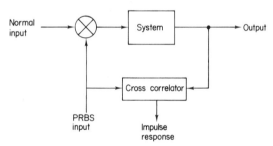

Fig. 6.15. General scheme for conducting a PRBS experiment.

The general scheme for conducting a PRBS experiment is illustrated in Fig. 6.15. Two types of behavioural experiment, conforming to this pattern, have been conducted in the author's laboratory. In one case, a digital computer was used to generate a PRBS signal, which turned the motor of a syringe driver on and off, in the situation illustrated in Fig. 6.16. When on, the microsyringe, filled with a chemical in aqueous solution, was driven at a constant rate calculated to provide an adequate dose over a period of 2 h. The chemical used was angiotensin II (Hypertensin CIBA), and the injections were made directly into the preoptic area of the rat brain. Injection of angiotensin into this site has been shown to have a marked effect on drinking in rats (Epstein *et al.*, 1970). The PRBS

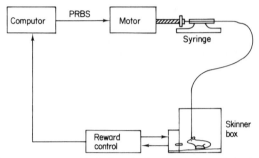

Fig. 6.16. Computer-controlled chemical stimulation of the brain. Intracranial injections are administered to a rat in a Skinner box in accordance with the PRBS signal. The effect of the injections on the rat's bar-pressing performance is recorded by the computer.

experiment is an attempt to determine the impulse response relating the injection to bar-pressing for water in a Skinner box. In order to minimize the effect of water rewards, and open the satiation feedback loop (see section 6.2.0), a VI (30-s), reward schedule was used. The main difficulty with this experiment is that the balance, between too great and too small a dose (i.e. between the effects of non-linearity and noise) is a delicate one. If the dose is too large the animal is unable to respond fast enough, the response measure saturates and the impulse response is non-linear. If the dose is too small, the animal does not respond sufficiently to provide the computer with the quantity of information needed to calculate the impulse response. Variation in the optimum dose level makes it very difficult to obtain reliable results. More promising results have been obtained from a study of the effect of angiotensin on feeding behaviour. The rats were food deprived and allowed to press a bar to obtain food on a VI reward schedule. The effect of the injection is to depress feeding, in line with the hypothesis that feeding is inhibited by thirst factors (McFarland, 1964; Oatley, 1967). The results of the PRBS experiment show that the impulse response calculated by the computer is similar to that obtained by measuring the transient depression of feeding following a single large injection of angiotensin (Fig. 6.17).

The second type of experiment is entirely behavioural and is based on the fact that *ad libitum* water intake is largely dictated by food intake (Fitzsimons and Le Magnen, 1969; McFarland, 1969a). The experiment is carried out on doves maintained continuously in a Skinner box under controlled climatic conditions. Water can be obtained *ad libitum*, 0.1 cm³ being delivered for each peck at the

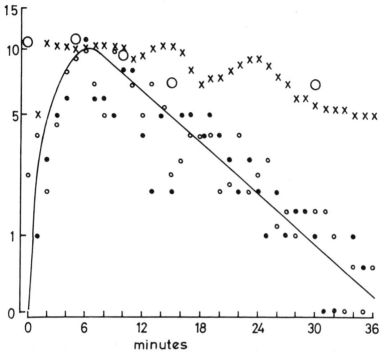

Fig. 6.17. Effect of intracranial angiotensin on operant feeding and drinking. (a) Depression of feeding in hungry rats following a single injection, giving data from two rats (open and closed small circles). Ordinate-control minus experimental score gives magnitude of depression in responses/min. (b) Depression of feeding following multiple, computer controlled (PRBS), injections. Solid l]ne indicates the calculated impulse response. (c) Stimulation of drinking in non-deprived rats following a single injection. Crosses give averaged results of the three rats tested (ordinate—responses/min). (d) Amount consumed in drinking tests conducted at various intervals (abscissa) following a single injection (ordinate—cc consumed), indicated by large circles. The effect of angiotensin upon drinking is concomitant with its decay of effectiveness at the site of injection, but the effect on feeding builds up more slowly and disappears more quickly, suggesting that the inhibition of eating is opposed by a reciprocal action on the part of the hunger control system. (From McFarland and Rolls, 1971.)

key. The availability of food is controlled by the PRBS (see Fig. 6.18), the parameters of which are arranged in such a way that the animal feeds most of the time that food is available. Thus feeding is regarded as the input to the system and drinking the output. Sample results are illustrated in Fig. 6.19. From these it can be seen that feeding has two types of effect on drinking. A direct effect,

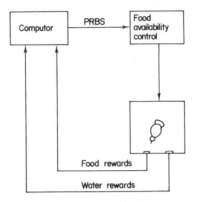

Fig. 6.18. Computer-controlled food availability in an otherwise free feeding and drinking situation. The dove pecks at the red and green illuminated keys to obtain food and water respectively. The computer calculates the relationship (impulse response) between food availability and water rewards.

occurring within 1 or 2 min, and a more delayed effect, which presumably reflects the change in systemic thirst factors following food intake.

In general, the use of PRBS methods in behavioural systems analysis seems promising, particularly in assessing the effects of one type of behaviour upon another. Although the PRBS method is subject to certain types of error, methods of allowing for these have been devised. As PRBS methods constitute an actively growing area in modern control theory, it is difficult to give an assessment of the capabilities of the method at this stage, and the reader should refer to the literature for further information (Korn, 1966; Williams and Clarke, 1968; Hughes, 1969).

6.2.3 Multiple-input-output Systems

Identification of a single transfer function or impulse response within a system, is only a small step in the analysis of systems as complex as those responsible for behaviour. In the analysis of multiple-input-output systems there are two definitive stages: (1) the demonstration that a causal relation exists between two variables, and (2) the determination of the dynamic nature of that relation. In behavioural studies the former problem has received considerable attention, while the latter has been rather neglected.

In a linear system interactions between variables are necessarily additive, in the sense that an operator H is linear if, and only if

$$H(x_1 + x_2) = Hx_1 + Hx_2. \tag{6.30}$$

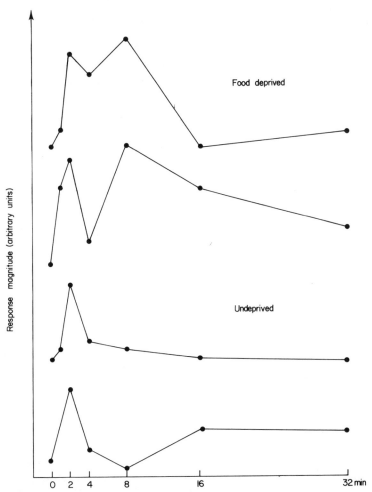

Fig. 6.19. Examples of impulse responses relating food availability and drinking behaviour. The peak at 2 min illustrates the well known close association between feeding and drinking (see section 5.2.3). The peak at 8 min in birds tested after food deprivation, is probably due to the fact that food is processed more quickly in hungry birds, and therefore causes systemic shifts in water balance. (From McFarland and Lloyd, 1971a.)

This is called the "superposition principle" (Brown, 1965). Thus the demonstration of additivity is an important step in determining the causal relations between variables, and this has been done in a number of cases. For example, additivity of cellular and extracellular stimuli for thirst in rats has been shown (Fitzsimons and Oatley, 1968; Fitzsimons, 1969; Blass and Fitzsimons, 1970), as has additivity of temperature and water deprivation on drinking in doves (Budgell, 1970) (Fig. 6.20). Additivity of external stimuli has been found in a number of cases and these are discussed in Chapter 7. The causal relations between variables demonstrated by studies of this type are essentially static in nature, and although important in the preliminary phases of systems analysis the rationale behind this type

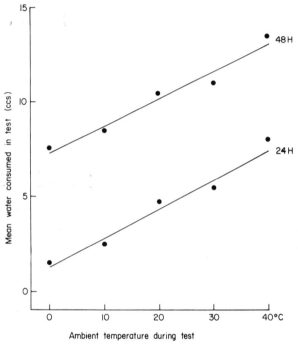

Fig. 6.20. Water consumption as a function of ambient temperature. Six doves were deprived of water for 24 or 48 h at 20° C. They were then transferred to a test-cage at a specified temperature and, after 30 min, were allowed to drink from a bowl. The amount drunk varied linearly with the temperature, the deprivation having an additive effect. (After Budgell, 1970.)

of approach is fairly commonplace in behavioural studies and is admirably documented by Hinde (1970).

The dynamic nature of interactions between variables in behavioural systems poses more of a problem. Delius (1969) has called for greater use of stochastic methods and has discussed many of the problems involved, and outlined some possible approaches. The present state of the art is delineated by Bendat and Piersol (1966), who suggest a combination of stochastic and matrix methods. The first step is to determine the cross spectral density function between each input and a selected output. In matrix notation an n-dimensional input vector is defined thus

$$[x](t) = [x_1(t), x_2(t), \ldots x_n(t)] \tag{6.31}$$

and an n-dimensional frequency response function vector

$$[H](\omega) = [H_1(\omega), H_2(\omega), \ldots H_n(\omega)] \tag{6.32}$$

next, an n-dimensional cross spectrum vector of the output $y(t)$ with the inputs $x_i(t)$

$$[\Phi_{xy}](\omega) = \Phi_{1y}(\omega), \Phi_{2y}(\omega), \ldots \Phi_{ny}(\omega) \tag{6.33}$$

where

$$\Phi_{iy}(\omega) = \Phi_{x_iy}(\omega) \qquad i = 1, 2, \ldots n$$

The matrix of cross-spectra between all the inputs is

$$\Phi_{xx}(\omega) = \begin{bmatrix} \Phi_{11}(\omega) & \Phi_{12}(\omega) \ldots \Phi_{1n}(\omega) \\ \Phi_{21}(\omega) & \Phi_{22}(\omega) \ldots \Phi_{2n}(\omega) \\ \cdot & \cdot \quad \ldots \\ \cdot & \cdot \quad \ldots \\ \Phi_{n1}(\omega) & \Phi_{n2}(\omega) \ldots \Phi_{nn}(\omega) \end{bmatrix} \tag{6.34}$$

The exercise moves, from here, into the realms of matrix algebra from which various interesting points are derived (Bendat and Piersol, 1966). Most notable is the possibility of determining each $H_i(\omega)$ as a function of the input-output cross spectra $\Phi_{xy}(\omega)$ and the input-output cross spectra $\Phi_{xx}(\omega)$, whether or not the various inputs are correlated. However, the determination of transfer functions between all possible combinations of input and output does not permit the internal structure of the black box to be determined uniquely. Behaviour alone cannot specify connections uniquely and the contents of the black box can be determined only with reference to independent knowledge concerning the nature of the system. The problem thus reverts eventually to the type of psychological problem discussed in the next chapter.

6.3 SIMULATION

In the development of systems analysis, the complexity of systems encountered has always outpaced the analytical means for their solution. This situation has led to the extensive development and use of physical aids to computation. Electronic computers are the most successful of these aids and may be classified into two main types, digital and analogue. Digital computers use an essentially numerical method of computation similar to that used by a human with a desk calculator. Analogue computers, as the name implies, provide an analogy with the system being studied. The analogy is based on the mathematical equivalence between the interdependence of the variables in the computer and that of the variables in the system.

6.3.1 Analogue Computers

Although digital computers are more powerful in their capacity to deal with complexity and with many different types of problem, analogue computers are particularly suited to the solution of differential equations in which time is the independent variable. They are also much cheaper and easier to use than digital computers. The programming of analogue computers is a similar process to systems analysis in general and their circuits have a particular affinity with block diagrams and flow graphs. The behaviour of any variable can be measured directly from a programme by attaching oscilloscope leads to the respective parts of the computer. Thus, they provide a simple way of checking block diagrams and flow graphs. But the most important use of analogue computers is the direct simulation of non-linear systems, for which no analytical theory exists.

A general purpose analogue computer basically consists of a number of "operational amplifiers" which are connected together to form a circuit, similar to the chains of cause-and-effect portrayed in block diagrams and flow graphs. All the system variables, other than time, are measured in terms of "voltage". An analogue computer programme is arranged so that voltage is manipulated in accordance with the relationship between the variables in the mathematical model. Independent variables are represented by input voltages, obtained from a suitable power supply, and dependent variables are derived from the input voltages, in accordance with the requirements of the model. Proportionality between dependent and independent variables can be obtained by altering the gain between the two variables. The gain can be reduced simply by inserting a potentiometer, as shown in Fig. 6.21a. To increase the gain above unity, it is necessary to use an operational amplifier. The gain of an amplifier is determined by the ratio between the values of the input and feedback resistances of the amplifier and can be represented by a small numerical label on the input to the amplifier, as shown in Fig. 6.21b. Unity gain is generally left unlabelled for simplicity. Note that operational amplifiers always reverse the sign of the input.

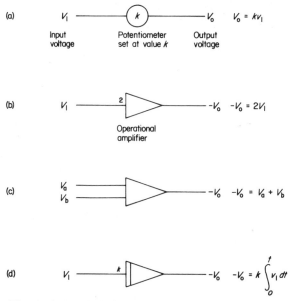

Fig. 6.21. Symbols for analogue computing elements.

In addition to gain determination and sign reversal, operational amplifiers can also be used for summation, in which case they are called "summing amplifiers" (Fig. 6.21c). Operational amplifiers may also be used to integrate input voltages with respect to time. Such an amplifier, called an "integrating amplifier", is shown in Fig. 6.21d, where k is a constant which determines the integral action time. By increasing or reducing k the time scale is expanded or contracted. Thus it is possible to specify the relationship between real time and computer time. It is also possible to differentiate a signal by using an operational amplifier but such an operation involves considerable practical difficulties and is usually avoided whenever possible. To summarize: operational amplifiers may be used to increase gain, change sign, add and integrate with respect to time. (For further details of analogue computing techniques see Hartley, 1962; Stewart and Atkinson, 1967.)

In programming an analogue computer it is usual to start from a set of questions, although it is also possible to set up an analogue computer circuit directly from a block diagram or a flow graph. Whichever method is used, it is important to specify exactly the mathematical significance of the transfer functions involved. Consider the system illustrated in Fig. 3.21. This can be simulated directly on an analogue computer, by joining the relevant amplifiers and potentiometers in the manner indicated in Fig. 6.22. Comparison

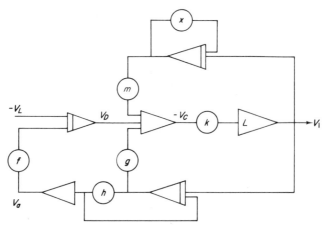

Fig. 6.22. Analogue computer circuit for Fig. 3.21 system. V_L = loss, V_D = debt, V_L = command, V_I = intake, V_A = absorption rate.

with Fig. 3.21 shows that the two representations of the system are very similar. The important variables of the system are represented directly by voltages at the output of amplifiers on the analogue computer and the parameters by potentiometer settings, thus permitting direct simulation of the system. This approach is useful primarily when the system contains non-linearities which are difficult or impossible to handle analytically. As an example, the simulation illustrated in Fig. 6.22 can be used to study the effect of limiting the ingestion rate V_I. This is done by means of a simple electronic circuit that limits the output of the amplifier L, so that it cannot go above a specified value. McFarland and McFarland (1968) compared the performance of this simulated system with that of animals whose

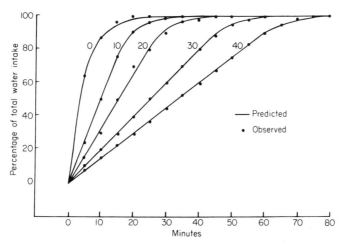

Fig. 6.23. Experimental limitation of the rate of operant water intake. Averaged results of five subjects at 0, 10, 20, 30 and 40 s time-out. (From McFarland and McFarland, 1968).

ingestion rate was limited experimentally, by introducing a time-out period following each reward in a Skinner box (Fig. 6.23).

Analogue computers have been used extensively in the simulation of physiological systems (see Grodins, 1963; Milhorn, 1966) and their use in behavioural systems analysis is growing.

6.3.2 Computer Simulation of an Homeostatic System

Although there have been a number of detailed studies of homeostatic systems in terms of control theory (Crosbie *et al.*, 1961; Grodins, 1959; Pace, 1961; Yates *et al.*, 1968), there has been relatively little work on behavioural aspects of homeostasis. McFarland (1965c, 1970c) and Oatley (1967) have touched upon some general aspects of problems involved in analysing thirst and drinking in terms of control theory. By virtue of its relative simplicity, the drinking control system is probably the most appropriate homeostatic system with which to start this type of behavioural analysis.

Water balance is maintained by control of intake and excretion of water and electrolytes. A considerable amount is known about the mechanisms responsible for control of excretion, rather less about intake, and very little about how the various sub-systems are integrated into a single complex homeostatic system. Time delays, non-linearities, and interactions between sub-systems and with other systems, make quantitative understanding of the regulatory processes involved almost impossible without a quantitative model. The procedure for obtaining such a model by computer simulation consists essentially in integrating the findings of many workers into a complex whole. Toates and Oatley (1970) have set out to develop such a model for the control system responsible for thirst and drinking in rats. Their work is by no means complete but sufficient progress has been made to illustrate some of the procedures used in computer simulation.

The simulation was developed in two main stages. Analogue simulations for each of the five major sub-systems were obtained and the values of important parameters found from empirical data. The whole system was then transferred to a digital computer and new predictions generated and checked against experiments. The sub-systems and their interactions are illustrated in Fig. 6.24, and Table 6.1 gives a list of the variables involved.

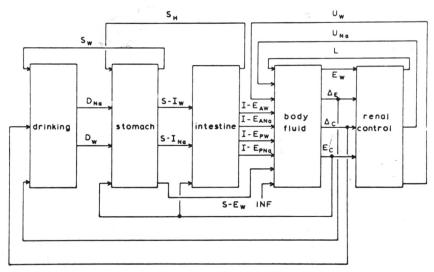

Fig. 6.24. Thirst sub-systems and their interconnections. See Table 6.1 for symbol key. (From Toates and Oatley, 1970.)

The sub-system responsible for interaction between the body fluid compartments (Fig. 6.25) is relatively straightforward. Body water is distributed between cellular and intracellular compartments and the respective potassium (C_c) and sodium (E_c) concentrations are indicated in the model. Strictly, the extracellular fluid is contained

Table 6.1

ADH	total ADH	$I-E_{P\text{Na}}$	passive flow of sodium
ADH_C	ADH concentration		between intestine and
ADH_E	effective ADH		ECF (m-equiv./min)
	concentration	$I-E_{PW}$	passive flow of water
ALD	total aldosterone		between intestine and
ALD_C	aldosterone concentration		ECF (ml/min)
C_C	cellular concentration of	K	potassium (m-equiv.)
	sodium in milli-equivalents	L	insensible water loss
	per ml (m-equiv./ml)		(ml/min)
C_W	cellular water (ml)	OS_W	osmotic flow of water
D_C	drinking fluid		between ECF and cells
	concentration of sodium		(ml/min)
	(m-equiv./ml)	S_C	stomach concentration
D_I	drinking inhibition factor		of sodium (m-equiv./ml)
	(ml/100 g body wt)	S_H	signal completely inhibiting
D_{Na}	rate of drinking sodium		stomach discharge
	(m-equiv./min)		(value 0 or 1)
D_W	rate of drinking water	S_I	stomach inhibition factor
	(ml/min)		(ml/100 g body wt)
Δ_C	cellular error (ml/100 g	S_{Na}	stomach sodium (m-equiv.)
	body wt)	S_W	stomach water (ml)
Δ_E	extracellular error	$S-E_W$	flow of water between
	(ml/100 g body wt)		stomach and ECF
ECF	extracellular fluid		(ml/min)
E_C	extracellular concentration	$S-I_{\text{Na}}$	flow of sodium from
	of sodium (m-equiv./ml)		stomach to intestine
E_W	extracellular water (ml)		(m-equiv./min)
GFR_N	glomerular filtration rate	$S-I_W$	flow of water from
	of sodium (m-equiv./min)		stomach to intestine
GFR_W	glomerular filtration rate		(ml/min)
	of water (ml/min)	T	total thirst signal
ICF	intracellular fluid		(ml/100 g body wt)
I.C.	initial conditions	$T_{F\text{Na}}$	tubular rejection factor
INF	infusion of sodium		of sodium
	(m-equiv./min)	T_{FW}	tubular rejection factor
I_C	intestine concentration		of water
	of sodium (m-equiv./ml)	TRF_{Na}	tubular reabsorption
I_{Na}	intestine sodium (m-equiv.)		factor of sodium
I_W	intestine water (ml)	TRF_W	tubular reabsorption
$I-E_{A\text{Na}}$	active flow of sodium from		factor of water
	intestine to ECF	U_C	urine concentration
	(m-equiv./min)		(m-equiv./ml)
$I-E_{AW}$	water carried by active	U_{Na}	urine sodium (m-equiv./min)
	sodium from intestine to	U'_{Na}	urine sodium for
	ECF (ml/min)		concentration check
			(m-equiv./min)
		U_W	urine water (ml/min)

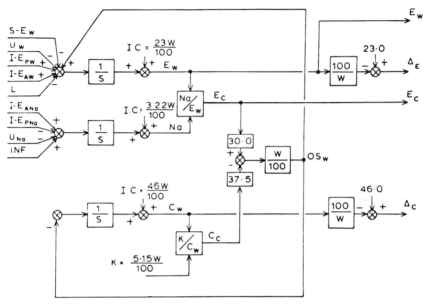

Fig. 6.25. Body fluid sub-system. (From Toates and Oatley, 1970.)

within two compartments, the vascular and interstitial, but the time-constant of the interaction between the two is sufficiently short for the distinction to be ignored in the present study. The potassium and sodium concentrations of the cellular and extracellular compartments respectively, are derived from a division of the respective volumes which introduces an essential non-linearity into the simulation. Subtraction of these gives the osmotic gradient (OS_w) which determines the rate of flow of water between the two compartments. The initial conditions and important variables in the model are calculated as deviations from their normal values per 100 g body weight.

In the model for the stomach (Fig. 6.26), the ingestion rates for water (D_w) and sodium (D_{Na}) are integrated separately and the two divided to give the concentration of sodium in the stomach. Stomach contents discharge into the intestine at a rate proportional to the contents remaining, the time-constant being a function of stomach concentration. The model also takes account of the osmotic flow between the stomach and the extracellular fluid which is partly dependent upon the degree of stomach distension (see Toates and Oatley, 1970, for details). The stomach model, then, is fairly complex and contains a number of non-linear components.

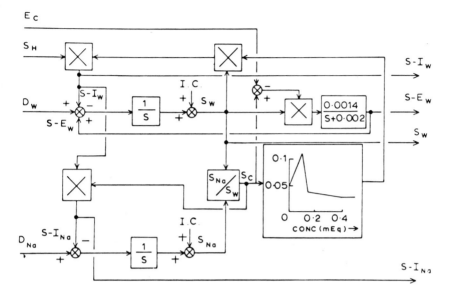

Fig. 6.26. Stomach sub-system. (From Toates and Oatley, 1970.)

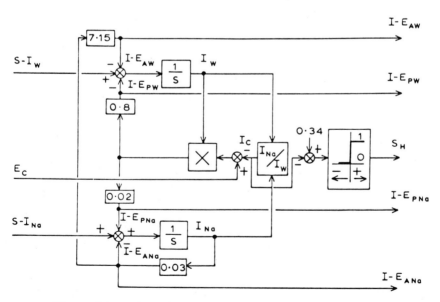

Fig. 6.27. Intestine sub-system. (From Toates and Oatley, 1970.)

The simulation of the intestine (Fig. 6.27) is basically similar to that of the stomach, the main additional features being the assumption that absorption depends upon the volume of fluid in the gut. In addition there is passive flow of water and active transport of sodium, between the intestine contents and the extracellular fluid. When the intestinal contents become hypertonic, stomach discharge is inhibited by a signal (S_H) having value 1 or 0. The behaviour of the model, following sudden stomach loads of hypotonic (Fig. 6.28)

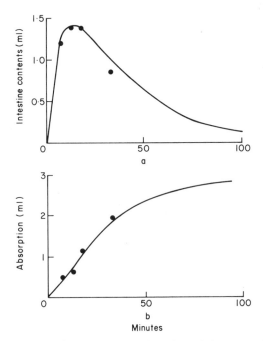

Fig. 6.28. Simulation of response to 3% body weight stomach load of 0.5% saline. (a) Intestine contents, (b) absorption of fluid from alimentary tract, Empirical results represented by points, simulation by continuous curve. (From Toates and Oatley, 1970.)

and hypertonic (Fig. 6.29) saline, shows quite good correspondence with experimental data.

The renal control of excretion of water and electrolytes is particularly complicated, and will not be dealt with here as it has limited direct behavioural relevance. The simulation by Toates and Oatley (Fig. 6.30) is based upon the work of a number of researchers, and takes account of the interactions between glomerular filtration rate and extracellular volume, water

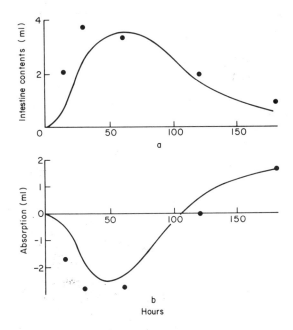

Fig. 6.29. Simulation of response to 3% body weight stomach load of 4% saline. Conventions as Fig. 6.28. Note initially negative absorption. (From Toates and Oatley, 1970.)

reabsorption and antidiuretic hormone concentration, and control of sodium excretion by aldosterone. Note that a number of non-linear processes are involved.

Oatley (1967) proposed that drinking is controlled by an additive combination of signals representing shrinkage of cellular and extracellular spaces, and there is now a considerable amount of evidence for such additivity (Fitzsimons and Oatley, 1968; Fitzsimons, 1969; Blass and Fitzsimons, 1970). In the model of the drinking sub-system (Fig. 6.31), changes in cellular (Δ_c) and extracellular (Δ_c) space are added. Expansion of the extracellular space seems to have no effect on drinking (Fitzsimons, 1963), and this feature is introduced as a non-linear "positive pass gate" which passes only positive values of Δ_c.

Drinking seems to be an all-or-none process in rats (Stellar and Hill, 1952; Corbit and Luschei, 1969), and this suggests that there are thresholds for the initiation and termination of drinking. Evidence for a threshold for the initiation of drinking is provided by the work of Wolf (1958) and Fitzsimons (1963), and such a

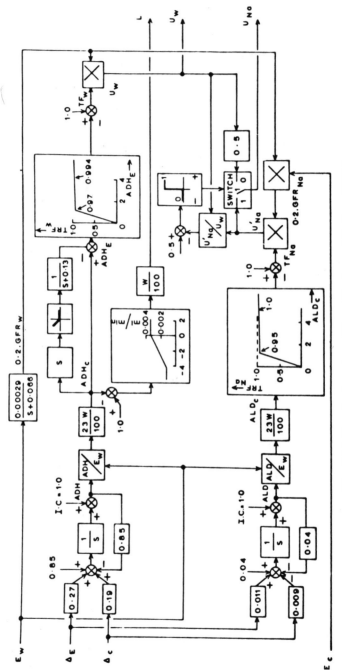

Fig. 6.30. Renal sub-system. (From Toates and Oatley, 1970.)

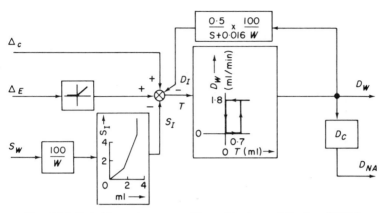

Fig. 6.31. Drinking sub-system. (From Toates and Oatley, 1970.)

threshold is included in the present model in the form of a hysteresis loop. The threshold is reached when the thirst signal T reaches a certain value and drinking ceases when this signal reaches zero. Because of the delay in absorption of water, it might be thought that the thirst signal is unlikely to reach zero while drinking is in progress, unless the animal is forced to drink very slowly (McFarland and McFarland, 1968). However, some work on rats (e.g. Novin, 1962) suggests that water absorption is sufficiently fast, and drinking sufficiently slow, to enable drinking to be terminated by this means. There is also evidence that stomach loading and oral factors can inhibit drinking (Adolph, 1950; Miller, Sampliner and Woodrow, 1957) and this possibility is indicated in the model by drinking (D_I) and stomach (S_I) inhibition factors which subtract from the excitatory effects of the systemic variables.

In conjunction with the simulation, Toates and Oatley (1970) carried out experiments involving intravenous saline infusions. In Fig. 6.32 the effect of continuous infusion of hypertonic saline in rats is compared with that of the model. The effect of the thirst threshold on the drinking response is clearly demonstrated in (a) and an interesting steady-state velocity error is evident in the sodium excretion response (b) and a similar effect occurs with water excretion (c). The drinking response to a single injection of hypertonic saline is illustrated in Fig. 6.33. The response of the model is faster than that of the rats, and the authors suggest that this is due to the animal pausing for intervals of several seconds and responding to competing motivations, such as grooming. This interpretation is open to question, because it has been shown that

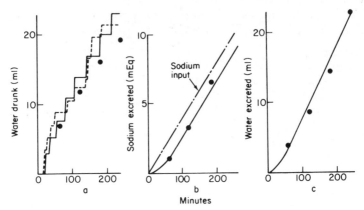

Fig. 6.32. Response of 430 g rat to continuous intravenous infusion at rate of 0.0405 mEq of Na/min. (a) Drinking, (b) sodium excreted, (c) water excreted as urine. Continuous curve–simulation, dotted curve–typical experimental result, points mean experimental result. (From Toates and Oatley, 1970.)

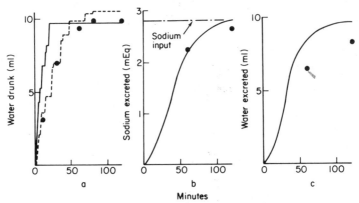

Fig. 6.33. Response of 430 g rat to step injection of 2.73 mEq of sodium. Conventions as Fig. 6.32. (From Toates and Oatley, 1970);

such interruptions of feeding and drinking in doves are under the control of the ongoing behaviour, being disinhibited and not arising by competition (McFarland, 1970b). This finding poses problems which will be discussed in the next chapter.

In general, the simulation can be said to be reasonably successful as far as it goes. That the system is of the correct order is shown by the correspondence of the observed and predicted steady-state errors in Fig. 6.32. Furthermore, the parameter values, based upon physiological values obtained from the literature, provide a remarkably good fit without any substantial adjustment being necessary. The limitations of the model lie chiefly on the behavioural side which takes no account of effects of temperature and food intake upon drinking, or of positive feedback from oral factors. But there is no reason why the simulation should not be extended to cover these and other aspects of thirst.

CHAPTER 7

Control Theory and Behaviour Theory

Control theory is concerned essentially with behaviour as a function of time. By contrast, behaviour as a function of time covers only part of the study of animal behaviour because psychologists are also concerned to discover how behaviour changes, as a function of generations, maturation and experience. Changes in behaviour as a function of time may be related to external stimulation, endogenous rhythms, physiological state and a number of other types of transient causal factor. Changes with time may also occur as a result of learning, maturation, or injury, but such changes are not generally reversible. When the response to the same external stimulation differs from one occasion to another, a change in internal state can be inferred. If the change is reversible, then a change in "motivational" state is involved. In this chapter it will be argued that the prime relevance of control theory to behaviour theory is in the area of motivation. This is not to exclude learning, maturation and similar phenomena, but to place motivation in a central position to which other behavioural phenomena may be related.

The essence of motivation is that changes in the internal state of the animal are related to changes in responsiveness to external stimuli. Thus a bird may incubate an egg at one time and eat it at another. In one case its motivational state is dominated by broodiness, in the other by hunger. Two considerations of prime importance from the viewpoint of control theory are the nature of the interaction of internal and external factors and the nature of the changes in internal state. This chapter is primarily devoted to these topics. A convenient starting point is Lorenz's model of a motivational system in Fig. 7.1

"the tap T supplying a constant flow of liquid represents the endogenous production of action-specific energy; the liquid accumulated in the reservoir R represents the amount of this energy which is at the disposal of the organism at a given moment, the elevation attained by its upper level corresponds, at an inverse ratio, to the momentary threshold of the reaction. The cone valve V represents the releasing mechanism, the inhibitory function of the higher centres being symbolised by the spring S. The scale-pan Sp which is connected with the valve-shaft by a string acting over a pulley represents the perceptual sector of the releasing mechanism, the weight applied corresponds to the impinging stimulation. This arrangement is a good symbol of how the internal accumulation of action-specific energy, and the external stimulation are both acting in the same direction, both tending to open the valve." (Lorenz, 1950.)

The outflow from the reservoir represents the motor activity of the behaviour, the intensity of which can be measured by the distance

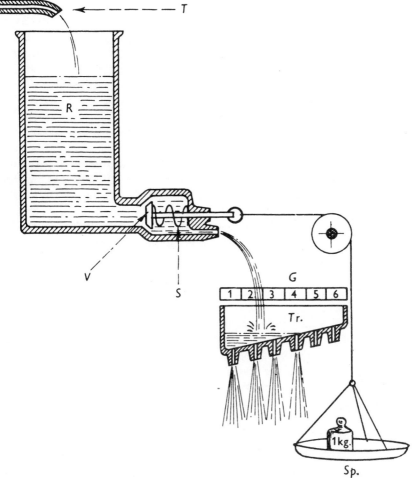

Fig. 7.1. Lorenz's hydraulic model for a motivational system. (From Lorenz, 1950.)

the jet reaches. The trough *Tr* symbolizes the recruitment of different components of the behaviour pattern as the intensity increases. The intensity of the observed behaviour is a joint function of the strengths of the internal and external causal factors.

7.1 INTERACTION OF EXTERNAL AND INTERNAL FACTORS CONTROLLING BEHAVIOUR

The necessity for distinguishing between internal and external factors controlling behaviour arises from the common observation that a given set of input conditions leads to different output phenomena, from one occasion to another. Although the inferred change in internal state must, ultimately, derive from external factors, it is useful to distinguish between short-term sensitizing and long-term organizing effects of external factors. Consideration of the development of song in birds may serve to illustrate this point.

Generalizations concerning the development of song patterns can be made on the basis of comparison of species differing markedly in performance (Konishi and Nottebohm, 1969). These generalizations may be tentatively summarized as follows: the naive bird has a "template" of the species-characteristic song which may, or may not, be modified by experience of the song produced by other individuals. Such modification may be restricted to a "sensitive period", which generally occurs before the bird itself begins to sing. The extent to which the template can be modified may be limited. In the chaffinch the limitation seems to be set by the resemblance to normal song, but in other species different mechanisms operate. In the second stage of song learning there appears to be a comparison between the template and the sound produced by the bird. Singing the "correct" song appears to be self-rewarding, for it reinforces the establishment of the song pattern in the animal's behavioural repertoire. The progressive improvement in vocalization, which marks this stage of song learning, does not occur in birds deprived of auditory feedback by deafening. However, the song pattern is little affected by deafening after the vocalization has reached its final form (Konishi and Nottebohm, 1969).

From the viewpoint of control theory, modification of a template by external stimulation, during a period before the bird itself begins to sing, involves changes in an essentially controllable but unobservable part of the system (section 6.1.2). The template is an hypothetical construct, whose existence is postulated on the basis of observations made long after the relevant manipulation of external stimuli. Moreover, in those species in which such external stimulation affects subsequent vocalization only during a specified sensitive period, the changes in the template are irreversible. Such changes in the organization of a system involve alteration of parameters and can be termed structural changes. These must be distinguished from changes of state resulting from the action of integrators within the system. Structural changes, like those involved in learning, maturation and injury, involve essentially non-linear processes, which in behavioural systems are generally irreversible. State changes, on the other hand, involve processes which are linear in principle, although non-linearities may be present in particular instances. These changes are generally reversible and in the behavioural context come under the general heading of motivational phenomena.

The role of the template in the second stage of song learning is partly uncontrollable, external stimulation having no further effect on the form of the template. Comparison between the template and the sound produced by the bird itself involves auditory feedback and presumably leads to reorganization of the song-producing mechanism. When the vocalization is fully developed, auditory feedback is no longer important and vocalization is probably controlled by proprioceptive feedback (Konishi and Nottebaum, 1969). At this stage the system responsible for vocalization is both controllable and observable, the template is relinquished, and the song may be said to have become fully incorporated into the animal's behavioural repertoire. The behaviour now occurs in response to appropriate combinations of internal and external stimuli. Not only can vocalization be influenced directly by external stimuli, such as territory signs, or the behaviour of other individuals (Mulligan and Olsen, 1969; Hooker and Hooker, 1969; Falls, 1969), but also indirectly, through the action of external stimuli upon the motivational state of the animal. Thus vocalizations and associated behaviour patterns can influence the hormonal state of other individuals (Lehrman, 1959; Brockway, 1969). Such influences are generally reversible, although hormonal influences during early development may have a permanent effect (Andrew, 1969).

The deliberately over-generalized example of avian vocalization used here is a paradigm for the types of problem associated with the general application of control theory to behaviour. External stimulation can have three types of effect upon a behavioural system: (1) it can induce structural changes, (2) alter the internal state of the system, and (3) have relatively rapid direct effects upon behaviour. The first of these effects is the most difficult to handle, as it involves essentially non-linear processes. Further investigation of this type of phenomenon is generally aimed at the question of how particular behaviour patterns become incorporated into the animal's repertoire, rather than how behaviour is controlled given a certain repertoire. The second two effects do not in principle provide any particular difficulty for control theory though, in practice, problems are likely to arise as a result of incomplete controllability and observability, and as a result of structural changes concurrent with state changes. The question of controllability and observability is partly one of isolating a suitable sub-system for study, as discussed in section 1.5. Questions of observability are also likely to arise in cases where the system itself forms a "model" of some part of its environment. This is discussed further in section 7.1.2. The problem of structural changes is similar to that of stationarity discussed in section 5.2.2. In the behavioural context, behaviour will be non-stationary if the system adapts by learning or maturation during the period under investigation. In motivational studies this effect can be reduced by employing animals whose performance is asymptotic with respect to learning, habituation etc. It may be possible sometimes to regard the motivational system as an adaptive control system (section 5.1.4), but progress in this area depends largely upon considerable knowledge of the motivational system at a stage in the adaptive process, which can be taken as a reference point. Studies of motivation and learning are complementary then, in the sense that the variables of a learning system appear as parameters in a motivational system.

To summarize the foregoing discussion in terms of Lorenz's hydraulic analogy, the motivational problem is essentially concerned with the dynamics of the liquid in the system and not with how the system came to have a particular structure. Although external factors may have been instrumental in determining the structure of the system, and in determining the level of liquid in the reservoir, their immediate relevance is in releasing the valve mechanism.

7.1.1 The Releasing Mechanism

Although it is generally agreed that behaviour is a joint function of internal state and external stimulation, there has been relatively little empirical research directed at the question of the nature of this function. For Hull (1943) the relation between internal and external factors is essentially multiplicative, $_sH_R \times D = {_sE_R}$. The generalized drive D is seen as multiplying habit strength $_sH_R$ to produce an excitatory potential $_sE_R$, which directly underlies the animal's performance. Habit strength, the tendency of a stimulus to evoke an associated response, is primarily a measure of the extent to which the animal is conditioned to the stimulus, but may also contain unconditioned elements. Hull (1952) also introduced the concepts of stimulus-intensity dynamism V, and incentive motivation K, as multipliers of drive, viz.

$$_sE_R = D \times V \times K \times {_sH_R}. \qquad (7.1)$$

Stimulus intensity dynamism is a measure of the perceived intensity of the stimulus. The greater the intensity, the greater the excitatory potential, for any given level of habit strength. The incentive motivation factor is a measure of the value of reinforcement, as anticipated by the animal. Incentive depends upon the learned associations of particular aspects of the external stimulus situation. The major determinant of excitatory potential is D, and the other multiplying terms in eqn. 7.1 are weighting factors each with a maximum of unity. Thus, when $_sH_R$, V and K are all maximal, $_sE_R = D$.

From the point of view of control theory, Hull's (1943, 1952) formulation has particular significance. His habit strength, incentive, and stimulus intensity dynamism, are all aspects of the external stimulus situation, as interpreted by the animal on the basis of past experience. They can be bracketed together as stimulus properties, and given the label S, viz.

$$_sE_R = D \times ({_sH_R} \times V \times K) = D \times S \qquad (7.2)$$

This formulation immediately suggests that S essentially fulfils the role of a parameter in a block diagram, as illustrated in Fig. 7.2. This

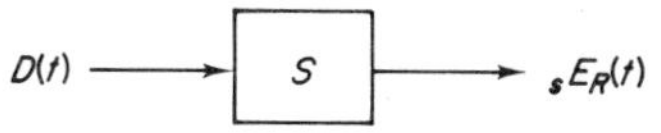

Fig. 7.2. Block diagram of Hullian formulation (eqn. 7.2).

parameter represents the properties of the environment as interpreted by the animal. Moreover, the fact that the value of S cannot exceed unity means that no amplification is involved, no extra energy is required, and the response can be said to be released rather than elicited by the external stimulus.

Hull's formulation is used here only as an illustration of how external stimuli can be incorporated into a control theory representation of a motivational system. For the purposes of long-term analysis, it is clear that environmental stimuli can usefully be considered as parameters through which motivational variables act. This is particularly so when the animal has been well trained and is thoroughly familiar with the experimental environment. The value of stimulus parameters can then be considered to be unity and the presence or absence of essential stimuli is equivalent to the open and closed states of a switch. For example, in the consideration of thirst satiation as a feedback process, outlined in section 3.3.1, the presence or absence of water is regarded as a switch at the level of the command to the ingestion mechanism. The latter thus receives a step-input when water is presented, the height of which is related to the length of prior deprivation. This type of formulation works well in practice, provided that the time-unit of analysis is not too short.

Some of the problems, associated with the role of external stimuli in more short-term analysis of motivational systems, can be illustrated by reference to Lorenz's (1950) concept of the releasing mechanism. As can be seen from Fig. 7.3, Lorenz's model implies that the release of "action-specific energy" depends upon not only the presence of appropriate external stimuli but also the level of liquid in the reservoir. Although there is a multiplicative relationship between the height H of the liquid, and the size of the aperture A (Fig. 7.3), the displacement of the valve itself (state variable X) is determined both by H, and by the force F exerted by the weight in the scale-pan. In other words, the "significance" of the external

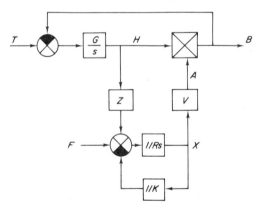

Fig. 7.3. Block diagram of Lorenzian motivational system. T = rate of inflow, B = rate of outflow, H = height of liquid in reservoir, F = force exerted by weight on scalepan, A = size of aperture, X = displacement of valve, K = spring constant, R = resistive friction of valve, G, Z, and V are parameters depending upon the geometry of the apparatus.

stimuli is influenced by the motivational state. Whereas the strict multiplication of the Hullian formulation implies that there can be no behaviour in the absence of appropriate external stimuli, Lorenz explicitly intended that sufficient pressure of the liquid alone may push open the valve, producing a "vacuum activity".

Lorenz's main justification for violating the straightforward multiplication principle is his observation that behaviour patterns for which there is strong internal motivation can occur "spontaneously", in the absence of appropriate external stimuli. Although quite well documented, the interpretation of such "vacuum activities" has been criticized on the grounds that it is never possible to be sure that no adequate external stimulation was present, or that the animal was not generalizing to stimuli not noticed by the observer (Bierens de Haan, 1947; Armstrong, 1950).

Lorenz's model can be interpreted as meaning that motivational state affects responsiveness to stimuli by altering the sensitivity of sense organs. As Hinde (1970) points out, the evidence is against such a view. Meyer (1952) found that 34 h of starvation produced no changes in the salt, sweet or bitter taste thresholds of human subjects, and electrophysiological studies provide similar evidence for rats (Pfaffman and Bare, 1950; Nachman and Pfaffman, 1963). Similarly, gonadectomy does not affect the ability of rats to distinguish sexually relevant odours (Le Magnen, 1952; Carr and Caul, 1962). Nevertheless, sodium deficient rats do show salt preferences, and intact male rats show preferences for odours characteristic of receptive females, while prepuberal or castrated males do not. Similarly, Delius (1968) found that colour preference in pigeons shifted after food or water deprivation. It appears that motivational state in some way affects the stimulus-response relation.

Another approach to the problem is to ask to what extent internal and external factors are independent. Bolles (1967) frames the question in Hullian terms, and reviews the evidence concerning the independence of D and H. The evidence consists mainly of various types of drive-shift experiment and studies of the effect of drive on stimulus generalization. Much of this work is more relevant to the question of whether what an animal learns depends upon its motivation during learning, than to the present problem. Bolles' (1967) conclusions however are interesting.

"It would appear that although there is some question in the case of avoidance behaviour the assumption of independence of D and H seems justified by the evidence for appetitive behaviour. However, even here we must add certain qualifications to Hull's original formulation. In a simple non-competitive situation, measures of response probability such as resistance to extinction and latency usually support the assumption of independence, and measures of response amplitude and vigor do not. In more complex learning situations the assumption of independence is apparently justified for a wide variety of performance measures. However, the evidence in this case does not provide substantial support for independence because it is not clear in highly response-competitive situations how D is supposed to combine with the various H's to determine performance."

Experiments more directly relevant to this issue have been carried out by ethologists. An example of the inverse relationship, between motivational state and the strength of external stimulus required to elicit a given response, is provided by the work of Baerends, Brouwer and Waterbolk (1955) on the courtship of the male guppy *Lebistes reticulatus*. The tendency of the male to attack, flee from, and behave sexually towards the female can be gauged from the colour patterns characteristic of each motivational state. In Fig. 7.4

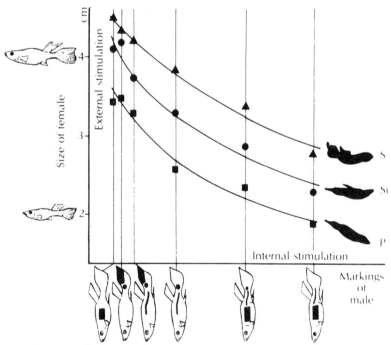

Fig. 7.4. The influence of the strength of external stimulation (measured by the size of the female) and the internal state (measured by the colour pattern of the male) in determining the courtship behaviour of male guppies. Each curve represents the combination of external stimulus and internal state producing posturing (*P*), sigmoid intention movements (*Si*), and the fully developed sigmoid (*S*), respectively. (From Baerends *et al.,* 1955.)

increasing sexual motivation is plotted as an index of colour change along the abscissa. The effectiveness of the female in eliciting courtship increases with her size and is plotted on the ordinate. The points plotted on the graph represent the relationship between the measures of internal state and external stimulation at which particular patterns of behaviour are observed. If the patterns P, Si, and S are taken to represent increasing values of response strength, and the scaling of the coordinates is taken at face value, then the curves obtained represent almost exactly those that would result from multiplication of internal and external factors. In practice, the methods of quantification are somewhat arbitrary, the scaling on the abscissa depending on the association of the different colour patterns with the relative frequency of activities characteristic of sexual tendency.

More quantitative data on the effectiveness of external stimuli in relation to motivational state is provided by Heiligenberg's study of attack readiness of the cichlid fish *Pelmatochromis subocellatus*. The strength of the attack readiness may be defined by the number of attacks it delivers per unit time in a standardized situation. Specifically

"An adult male fish is placed together with a group of young fish for several weeks. The male can attack the young fish at will. However the young fish always escape before being seriously bitten, so that a real fight—which might exhaust the male—never occurs. The behaviour of the male is recorded for 15 min; then a dummy of another male is presented behind a glass pane for half a minute, and the behaviour of the male is again recorded for the next 30 min. During the presentation of the dummy the male watches it, very rarely doing anything else than standing quietly in his place. Immediately after the removal of the dummy the male attacks the young fish much more than before and then returns slowly to his previous level of aggression." (Heiligenberg, 1965.)

Because the attack rate of the fish fluctuates considerably it is impossible to predict, from knowledge of its value in a given time interval, the exact rate for the subsequent interval. However, the expected value y, of attack rate within a specified interval, can be estimated as a function of the observed attack rate x. The attack rates observed within pairs of successive standardized time intervals, separated by a short intervening interval, were separated into classes on the basis of the level of attack observed in the first interval. Within each class the average values and variances were calculated for the attack rate within the first interval x and within the following interval y. As can be seen from Fig. 7.5, the relationship between the two is linear. The same procedure is applied when a dummy is

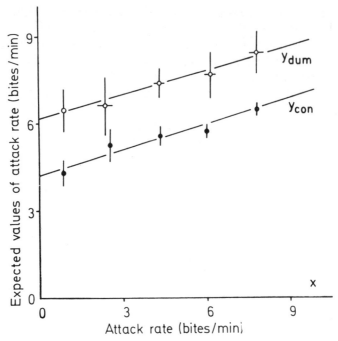

Fig. 7.5. Expected values of attack rate y_{con} (ordinate) within a subsequent 400-s interval as a function of the attack rate x (abscissa) observed with a given 400-interval (black circles). The expected value y_{dum} is additively increased if a dummy is presented between the given and subsequent 400-s interval (open circles). Data were divided into classes according to different levels of attack rate within the first 400-s interval. Within each group, average values (circles) and variances of the mean (bars) were calculated with respect to both coordinates. (From Leong, 1969.)

presented in the short intervening interval. Presentation of the dummy always altered the relationship between x and y by a constant amount (Fig. 7.5). In other words, the increment in attack rate caused by presentation of a dummy is independent of the pre-stimulatory attack rate and the presentation is additive in its effect on the stimuli already existing.

Using Heiligenberg's method with the cichlid fish *Haplochromis burtoni*, Leong (1969) found that different components of the colour patterns of territorial males, painted on dummies, were additive in their effects upon attack rate. This finding is in agreement with the "law of heterogeneous summation", first propounded by Seitz (1940). Similar findings come from other ethological studies. For example, Baerends (1962) analysed the stimulus components of eggs which elicit the egg retrieving behaviour in the herring gull and found that the effects of some components were additive in the mathematical sense.

From the point of view of the application of control theory to behaviour, two important points arise out of these ethological studies of the "releasing mechanism". The first point is that the additive effects of external stimuli upon response strength are clearly contrary to both the Hullian and the Lorenzian formulations. There is no particular significance in this fact, except that it demonstrates the danger of making assumptions, however attractive, without thorough empirical verification. To postulate a multiplicative relationship between motivational state and external factors, though adequate for some purposes, can only be regarded as an oversimplification. Indeed, it is unlikely that a postulated unitary interaction can ever stand up to close scrutiny since, in addition to their "releasing" function, external stimuli are also important in orienting the response and in altering motivational state (Hinde, 1970).

The second point is that it is possible to specify quantitatively, by thorough experimental analysis, the relationship between internal and external factors in the determination of behaviour. Although the analysis of behavioural systems in terms of control theory can make considerable progress for situations in which the external situation is held constant, consideration of the effects of changing external stimulation must be an essential part of future research into the short-term control of behaviour. Though little work of this type has been done so far, Heiligenberg's approach provides a good starting point.

Using a similar method to that employed with fish, Heiligenberg (1966) found that increase in chirp rate in a territorial male cricket *Acheta domesticus,* induced by playing chirps from a tape recorder, is independent of the cricket's chirp rate before stimulation, as long as the highest possible chirp rates are not reached. But the size of the increase depends upon the chirp rate of the stimulus used (Fig. 7.6). After the end of stimulation, the chirp rate returns exponentially to its previous level. In subsequent experiments Heiligenberg (1969) found that about 99% of the additive increment in chirp rate decays with a short time-constant (phasic effect), the rest decaying with a time-constant about 70 times as long (tonic effect). The effects of successive stimulus chirps are superimposed, but neither phasic nor tonic total increments ever pass a particular upper limit. If the total tonic increment becomes sufficiently large, a previously silent cricket

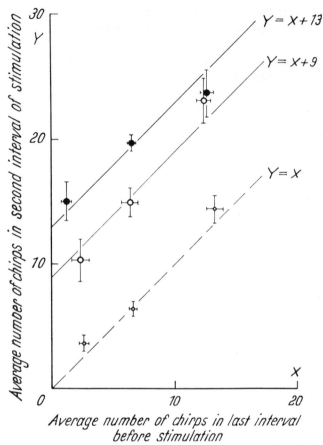

Fig. 7.6. Relation between number of chirps given by a cricket in the last 15-s interval before stimulation and the first two 15-s intervals after stimulation. Effects of stimulation rate of one chirp per 0.625 s (black circles), and 2.5 s (large open circles), are additive to those with no stimulation (small open circles). (From Heiligenberg, 1966.)

will start to chirp, and if the stimulus chirps are played at a sufficiently high rate the cricket tries to chirp in alternation. All but the last phenomenon were represented in a mathematical model (Fig. 7.7), which was then tested on an analogue computer. The model responded quantitatively as a cricket to all patterns of stimulation. Alternate chirping resulted directly from the model and required no special explanation.

In both the fish experiments, and the cricket experiments, presentation of an external stimulus produces an increase in response rate, which decays exponentially to its previous level when the stimulus is removed. This type of result shows that the effect of the stimulus is temporarily stored and in the case of the cricket experiments this observation provides the starting point for further analysis. Thus storage processes in a behavioural system can be identified by manipulation of external stimuli as well as by interruption of ongoing behaviour. Numerous cases exist in the literature (see for example Fig. 6.3), but although providing essential clues to the organization of the control system involved, they have rarely been followed up.

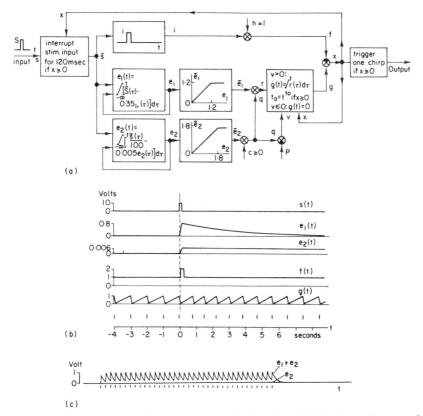

Fig. 7.7. Flow diagram (a) of a mathematical model of chirping and some of its time functions (b) with a single stimulus chirp being set at time $t = 0$. Let $v > 0$: As soon as

$$g(t) = \int_{t_0}^{t} r(\tau) \, d\tau \text{ (time } t \geqq t_0, \, r(\tau) \geqq 0)$$

(Caption cont. on p. 194)

7.1.2 Reafference

Interactions .between external and internal factors controlling behaviour can involve closed-loop phenomena, in addition to the open-loop type of interaction discussed in the previous section. Interaction between two sets of causal factors may be called open-loop when the factors are independent of each other. The

---

reaches the value f, $x = g - f$ is no longer negative. At this moment $\hat{t}$, a chirp will be triggered and $g(t)$ will be reset to zero with $\hat{t}$ becoming the new value of t_0. From now on $g(t)$ again starts to rise and the cycle is repeated. With $f = 1$ and only slow fluctuations of r, the frequency of the relaxation process $g(t)$ is approximately equal to r, i.e. the chirp rate of the model is proportional to r and thus an additive change in r will result in an additive change in chirp rate. If no stimulus s has been given for a sufficiently long period of time, i, e_1, e_2 and hence $\tilde{e}_1$, $\tilde{e}_2$ are zero, then $r = \tilde{e}_1 + q = \tilde{e}_1 + (\tilde{e}_2 + c) = c$, which is the spontaneous chirp rate of the model. If a stimulus does not coincide with a chirp $\tilde{s}$ will be identical to s, causing an inhibitory function i added to f, and two excitatory functions, namely the phasic function e_1 and the tonic function e_2, added to r. The function i equals 1 from the 40th to the 180th ms following the onset of the stimulus and is zero otherwise. This brief elevation of f prevents x from becoming positive within this particular period of time, so that no chirp will be triggered. $\tilde{e}_1$ and $\tilde{e}_2$, equal to e_1 and e_2 but never larger than 1.2 and 1.8 respectively, additively increase the chirp rate r. Both exponentially vanish again with half life periods of 2 and 140 s, i.e. time constants of 0.35 and 0.005, respectively.† If p, which represents behaviour activities antagonistic to chirping, is larger than $q = c + \tilde{e}_2$, r will be negative and hence $g(t)$ will be constantly zero. The silent model then can be made to chirp again by repeated stimulation so that with a sufficiently large value of e_2, q will become larger than p. In Fig. b the value of h was for simplicity assumed to be constantly equal to 1, so that $f(t) = h(t) + i(t)$ also equals one, except for a short period of time following stimulation, when $i(t) \neq 0$. However, in the model itself h is a Gaussian random variable with mean = 1 and variance = 2.5. h changes its value about three times in a second. The occurrences of chirps are indicated by vertical bars. In Fig. c the values of e_2 and of $e_1 + e_2$ have been plotted over time t with 36 stimulus chirps being presented at 5-s intervals (vertical bars). No stimulus coincided with a chirp, making it ineffective. Since e_1 vanishes rapidly, the value of $e_1 + e_2$ is equal to e_2 after a sufficiently long period of time after stimulation. At low stimulus rates, as for example in this figure, e_1 and e_2 never reach values higher than 1.2 or 1.8 respectively, so that $\tilde{e}_1$ and $\tilde{e}_2$ are always identical to e_1 and e_2.

† The formula

$$e(t) = \int_{\infty}^{t} (a \cdot s(\tau) - e(\tau) \cdot b)\, d\tau$$

has to be interpreted as follows: Up to the present time t all stimulus inputs $s(\tau)(-\infty < \tau < t)$, starting at τ_1, τ_2, τ_3, ... have been integrated after being weighted with the factor a. However, the value e of the integral continuously has been decreased by an amount of $e \cdot b\,d\tau$, which results in an exponential decay process following each single stimulus. (From Heiligenberg, 1969.)

possibility that stimulus significance, stimulus preferences, etc. can be influenced by motivational state has already been discussed (see section 7.1.1), and any hypothesis in which stimulus filtering is influenced by motivational state, or its behavioural consequences, is essentially a closed-loop type of hypothesis. Von Holst and Mittelstaedt (1950) (see von Holst, 1954) distinguish between "exafferent" stimulation, which is unaffected by internal factors, and "reafferent" stimulation, which occurs as a result of bodily movements. An animal capable of orienting itself must be capable of distinguishing between reafferent and exafferent information. Von Holst and Mittelstaedt (1950) proposed that animals do this by making use of information from the motor control centres and their theory has come to be called "reafference theory". Specifically, the motor commands not only actuate the muscles but also set up an "efference copy" which corresponds to the "expected" input from the exteroreceptors, in consequence of bodily movement. A comparison is made between the actual afferent stimulation and that which should have occurred according to the momentary motivational situation (see Fig. 7.8). When the afferent stimulation is composed entirely of reafferent signals the result of the comparison is zero, and when exafferent information is present it is detected by the comparator. The consequences of this comparison vary in accordance with the nature of the system. They may serve to direct

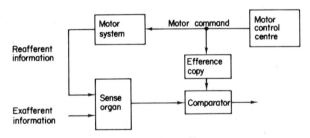

Fig. 7.8. Outline of basic reafference system.

subsequent bodily movement or to provide perceptual information concerning exafferent stimulation.

Von Holst and Mittelstaedt (1950) showed that the fly *Eristalis,* placed inside a cylinder painted with vertical stripes, shows a typical optomotor reflex, turning in the direction of the stripes when the cyliner is rotated. Such reflexes do not occur when the fly moves of its own accord, although the visual stimulation is similar. When the head of the fly is rotated through 180° (Fig. 7.9), the optomotor

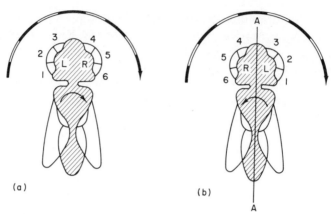

Fig. 7.9. The insect *Eristalis* in a rotating cylinder. L, R, left and right eyes. A, head in normal position. B, head rotated through 180°. (From von Holst, 1954.)

response is reversed as expected, but when the fly attempts to move of its own accord its movements appear self-exciting and it tends to go into a spin. These results show that the fly distinguishes between exafferent and reafferent movement stimuli and does not merely block the optomotor reflex during self initiated movement. Von Holst and Mittelstaedt (1950) invoke reafference theory to account for this phenomenon, and postulate that output from the comparator (Fig. 7.8) induces movement to the right or left according to the sign of the output. When the output is zero movement ceases and the target condition is achieved. When the head of the fly is reversed the sign of the reafference signal is reversed, resulting in a positive feedback situation. The unstable spinning behaviour is thus a consequence of the fact that the comparator output is able to determine bodily movements.

Von Holst also postulated that the output of the comparator could be responsible for perception of movement

"Consider my eye mechanically fixed and the muscle receptors narcotised (Fig. 7.10a). When I want to turn my eye to the right, an efference E and, according to the theory, an efference-copy EC is produced, but the immovable eye does not produce any reafference. The efference-copy will not be nullified, but transmitted to higher centres and could produce a perception. It is possible to predict the exact form of this perception (von Holst and Mittelstaedt, 1950). The perception, if I want to turn my eye to the right, must be that 'the surrounds have jumped to the right'. This is indeed the case! It has been known for many years from people with paralysed eye muscles and it has been established exactly from the experiments of Kornmuller on himself that every intended but unfulfilled eye movement results in the perception of a quantitative movement of the surroundings in the same direction. Since here *nothing* happens in the afferent pathways, this false perception *can* only result from the activity, originated by the intention of the eye movement, being returned to higher centres. This is another way of saying that the unmatched efference-copy causes the perception."

Von Holst (1954) went on to describe how the perception could be cancelled by appropriate mechanical movement of the eyeball, in the narcotized preparation (Figs 7.10b and c).

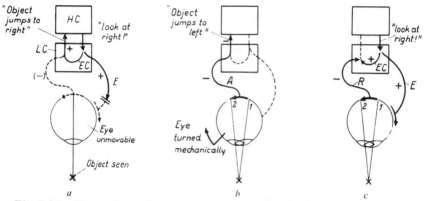

Fig. 7.10. Illustration of von Holst's experiments with the human eye. *HC* = higher centre, *LC* = lower centre, *EC* = efference copy, *E* = efferent message, *A* = afferent message, *R* = reafferent message. (a) Illustrates the situation when the eye is narcotized (see text), (b) the situation when the eye is moved mechanically, (c) represents situations (a) and (b) combined in such a way that the two effects cancel each other. (From von Holst, 1954.)

Von Holst's (1954) view is similar, though more precise, to the "outflow theory" of Helmholtz (1867), and can be contrasted with the "inflow theory" of Sherrington (1918) in which proprioceptive feedback from the extraocular muscles plays an important role (Fig. 7.11). Reafference theory is essentially a development of the concept of open-loop control. It is of functional importance in cases where rapid control or decision is at a premium, and it is feasible in cases where the consequences of the behaviour are not normally subject to disturbance and are therefore predictable. The eyeball, for example, is not normally a loaded organ and movement instructions generally have predictable visual consequences (see section 2.1). Reafference theory involves a type of predictive control, because the efference copy can be regarded as a representation of the expected consequences of the behaviour. This feature of the theory has led a number of workers to apply reafference theory to situations not envisaged by its original proponents.

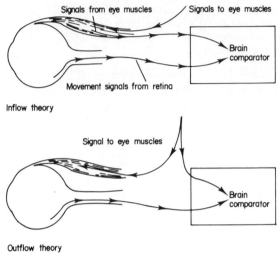

Fig. 7.11. Inflow and outflow theories of eye movement control. (From Gregory, 1966.)

Reafference theory has been applied to visual-motor adaptation, of the type illustrated in Fig. 7.12 (Held, 1961; Hein and Held, 1962). In terms of the scheme illustrated in Fig. 7.13

"... the reafferent visual signal is compared (in the Comparator) with a signal selected from the Correlation Storage by the monitored efferent signal. The Correlation Storage acts as a kind of memory which retains traces of previous combinations of concurrent efferent and reafferent signals. The currently monitored efferent signal is presumed to select the trace combination containing the identical efferent part and to activate the reafferent trace combined with it. The resulting revived reafferent signal is sent to the Comparator for comparison with the current reafferent signal. The outcome of this comparison determines further performance" (Held, 1961).

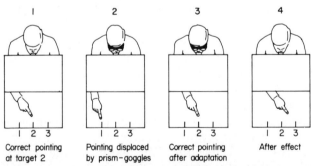

Fig. 7.12. Four stages of adaptation to laterally displaced visual field. (From McFarland and McFarland, 1969; after Rock and Harris, 1967.)

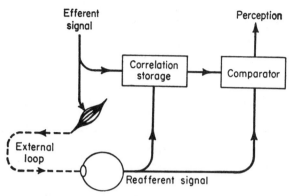

Fig. 7.13. Schematized process assumed to underlie the consequences of rearrangement, neonatal development, disarrangement and deprivation on visual-motor coordination. (From Hein and Held, 1962.)

Held and his co-workers compared the effectiveness of self-produced movements with that of passive movements in the readaptation of visual motor coordination. They claimed that only the active movement condition led to any significant readaptation (Held and Hein, 1958; Held and Freedman, 1963), and (Hein and Held, 1962) that reafference was necessary for such a change to take place. Held and Hein (1963) carried out experiments in which an active kitten was linked mechanically to a restrained passive kitten (Fig. 7.14).

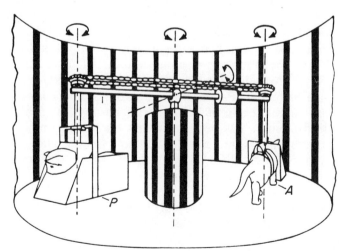

Fig. 7.14. Apparatus for equating motion and consequent visual feedback for an actively moving animal (A) and a passively moving animal (P). (From Held and Hein, 1963.)

Both had similar visual experience but only in the active kitten could this be linked to self-produced movement. Only the active kitten developed the ability to avoid a visual cliff, blink at an approaching object and extend its paws to a surface. In another experiment, each animal in turn was active and passive, but with one eye open under one condition and the other eye open in the other. The animals could perform the three tests only when the "active" eye was open. Held and Hein concluded that reafferent stimulation is essential for the development of visual motor coordination. This view has been contested both on empirical grounds and on methodological grounds (Howard and Templeton, 1966). The main criticism is that unless the subject is given relevant exafferent stimulation in the control condition, it can never have the opportunity to learn. Howard, Craske and Templeton (1965) showed that exafferent information regarding optical distortion can lead to some visual-motor adaptation in passive human subjects. Thus, while it is probable that reafference plays some role in visual-motor adaptation, its exact nature remains in question.

Another area of study in which a reafference-like principle has been invoked is that concerned with an animal's response to novelty. In response to a novel stimulus many animals show a combination of somatic and autonomic responses, known as the "orienting response". If the stimulus is repeated without significant consequences to the animal, the response wanes. The waning is fairly stimulus specific, and the response may reappear if the characteristics of the stimulus are changed. In accounting for the results of his considerable investigations of this phenomenon, Sokolov (1960) suggested that the orienting response appears whenever the sensory input does not coincide with a "neuronal model" which is set up as a consequence of repeated stimulation. Sokolov's theory is concordant with a considerable body of behavioural and physiological evidence (Hinde, 1970) but it has been criticized on the grounds that it is more elaborate than is necessary to account for the evidence (Horn, 1967). In particular, Horn (1967) claims that the concept of a neuronal model is an unnecessary one and that the phenomena can be accounted for by "self-generated depression" within the responding system. This controversy, which is complicated by the many factors involved (see Hinde, 1970, for an excellent review), illustrates the more general problem of the criteria that should be taken into account in incorporating concepts such as "efference copy" or "neuronal model" into a hypothetical model or system.

In many respects the problem is similar to that of the

identification of a "set point", discussed in section 2.2. For any proposed model there may always be an alternative which will have the same performance characteristics. The alternative may be different from the original in a small respect but this difference may be sufficient to nullify the exclusiveness of a "set point" or "efference copy". In short, where physical identification is not possible it may not always be meaningful to insist on the importance of a particular concept, unless it can be shown to fulfil a unique role in the explanation of the phenomenon in question.

As an example, consider the use of "efference copy" concepts in motivational studies. Bastock, Morris and Moynihan (1953) suggested that, at the initiation of an activity, a "centre" is charged by an "output copy" (efference copy) of the "normally expected" stimuli. Failure in sensory feedback (i.e. in confirmation of the expectation) was supposed to lead to an accumulation of nervous "energy" that could be responsible for displacement activity. This idea was taken up by McFarland (1966b), who replaced the notion of "accumulation of energy" with the proposal that a feedback discrepancy, resulting from a mismatch within the comparator (Fig. 7.15), would cause the animal to switch attention and so act as

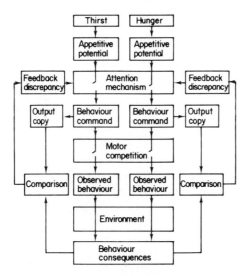

Fig. 7.15. McFarland's explanation of displacement activities in terms of a reafference theory. Disruption of behaviour consequences by motor competition, or environmental factors results in a feedback discrepancy which induces a switch of attention from the cues controlling the ongoing behaviour. (From McFarland, 1966b.)

a mechanism of disinhibition. In other words, any disruption of ongoing behaviour, such that the "expected" consequences are not confirmed, will cause attention to be diverted from the stimuli controlling that behaviour. In the long term, such switches in attention will effectively reduce the general motivational level (i.e. averaging across periods of activity and non-activation of the behaviour command), and the attention mechanism (Fig. 7.15) can be represented as a summing point in a conventional block diagram portrayal of the fundamental part of McFarland's theory (Fig. 7.16).

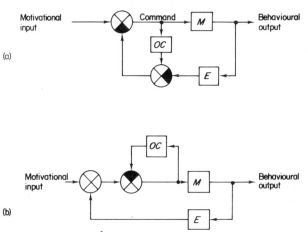

Fig. 7.16. Alternative (b) interpretation of motivational reafference theory (a). OC = output copy, M = motor control system, E = environmental situation.

From the simplified representation of a motivational reafference mechanism illustrated in Fig. 7.16a it can be seen that OC, the "output copy", does not play an essential role in the system. For example, in the version illustrated in Fig. 7.16b which is functionally identical to Fig. 7.16a, OC appears as a parameter in a simple feedback loop. This alternative version of the theory states that the command wanes unless positive feedback from the consequences of behaviour are available to cancel this self-depression. While, in reafference theory, the nature of the output copy is determined by the learned relationship between the behaviour and its consequences (McFarland, 1966b), in the alternative version it is the characteristics of the self-depression that are altered by learning. At the behavioural level, it is difficult to see how it would be possible to distinguish the two alternative mechanisms and the justifiability of the "output copy" concept must therefore be questioned.

Bastock, Morris and Moynihan (1953) invoked the concept of an "output copy" to explain a bird's reaction to a deficient nest situation, and this view has since been supported by Beer (1961, 1965) and by Baerends (1970). Although the latter study involves a number of complications (Fig. 7.17), in essence it is similar to that outlined above. Baerends (1970) proposes that "the incoming feed-back information from sitting on the clutch is checked against a preset stimulus expectancy value that is derived from an output copy connected with the command to the muscles to perform incubation". When the feedback exceeds the "expected" the tendency to incubate is increased, and when it falls short the probability of incubation being interrupted is increased. Baerends (1970) maintains that the majority of the interruptive behaviour (building, preening, etc.) is disinhibited following "looking around", and in this respect his general theory is similar to that of McFarland (1966b).

Baerends' (1970) study raises the possibility of an alternative explanation of the type illustrated in Fig. 7.16b. According to Baerends

"The crucial argument for assuming the existence of some kind of preset template for measuring the feed-back input is, that when varying experimentally the number, the temperature, and the size of the models in the nest bowl, the interruptive behaviour turned out to be at least an optimum of three eggs, of 37-39° C internal egg temperature and of normal size and shape."

It is also possible that such optimality could be achieved by means of a non-linear filter in E, the relevant condition of the environment as measured and interpreted by the animal. For instance, the output of E could be low when there is one egg in the nest, high with three, but low with five eggs in the nest. On this basis either of the alternatives illustrated in Fig. 7.16 could underlie the behaviour observed by Baerends. In other words, it is possible that the tendency to incubate is self-reducing unless adequate positive feedback is available from the eggs and nest. This type of explanation has its attractions, in that it is concordant with the Hullian principle of reactive inhibition according to which performance of any activity is accompanied by self-inhibitory effects. It is also consistent with the view that performance of an activity may result in positive feedback, and thus be self-reinforcing (section 3.3.3).

A more telling observation in favour of Baerends' (1970) theory is the observation that "the responsiveness to inadequate feed-back stimulation from the nest-bowl, as measured by interruptive

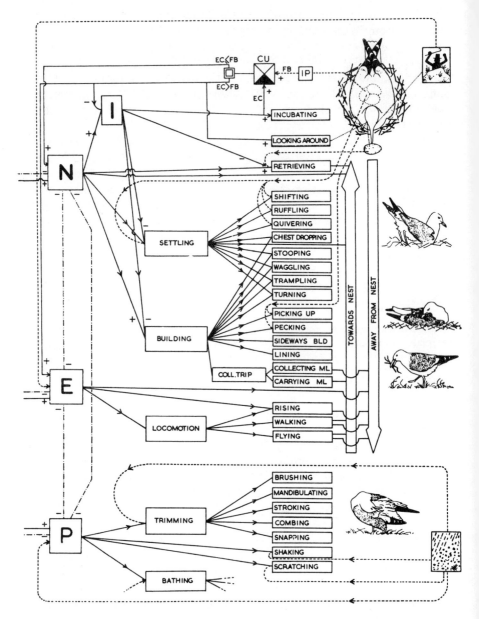

Fig. 7.17. Outline of the factors controlling incubation behaviour of the Herring gull. *EC* = efference copy, *CU* = comparison unit, *FB* = feedback stimulation from the eggs. (From Baerends, 1970.)

behaviour, is low when the tendency to incubate is low, and increases with a rise in this tendency". Baerends goes on to observe that

"Our model has to account for this fact. This can be done by assuming that the value of the efference copy (*EC*) is not constant, but depends on the output of *I*, the incubation tendency. As a result, the total feed-back input in the comparing unit (*CU*), in order to match the efference copy, has to be greater (more complete) at high than at low tendencies to incubate. Consequently the bird will be more sensitive to deficiencies of the nest situation when its incubation tendency is high."

It is more difficult to account for this phenomenon in terms of the self-waning theory (Fig. 7.16b). On this view, the time-constant of decline in observed behaviour, which is proportional to the amount of interruptive behaviour observed, is dependent on the parameters of the parallel feedback loops (i.e. in practice, to the inadequacies of *E*). An increase in motivational input would alter only the asymptotic value of the waning function which should make no difference to the amount of interruptive behaviour observed. Of course, no such definite predictions can be made with confidence on the basis of such loosely formulated models. Much depends upon the dynamics of a model and only a great deal of further research can hope to settle this issue.

The main difficulty in assessing the validity of reafference theories is that questions of observability arise in cases where the system forms a "model" with respect to some part of its environment (see section 7.1.0). In the case of von Holst's (1954) eye-movement example, this difficulty is partly overcome by the use of the subject's reports concerning his perceptions. By this means von Holst is able to "open the black box" and obtain information concerning a critical point in the system. In the case of animal work, where only overt behavioural measures are available, the problem of identification of the "output copy" is more difficult and cannot yet be said to have been convincingly demonstrated.

7.2 MOTIVATIONAL SYSTEMS

A motivational system can be defined on the basis of the assumption that the total behaviour repertoire of an animal is composed of a number of distinct categories of behaviour, each of which is controlled by one of a number of separate systems, called "instincts" by Tinbergen (1951) and more recently referred to as motivational systems. Although this assumption has been challenged in the past, the specificity of "drives" now seems to be generally accepted as a working hypothesis (see section 7.2.3).

A motivational system can thus be envisaged as a system controlling a group of functionally related activities. It is convenient to refer to such systems in general terms, as "feeding system" or "aggression system". The variables of a motivational system relate to behaviour as a function of time, examples being "hunger", "frustration" and other typically "motivational" variables. The parameters of a motivational system relate to a wide variety of mechanisms, such as sense organs, calibration mechanisms, accumulative processes (integrators), and mechanisms concerned with weighting the significance of external stimuli. As discussed in section 7.1.0 the values of such parameters undergo alterations due to learning, maturation and changes in the external stimulus situation. These changes complicate the process of analysis by introducing non-linear and non-stationary phenomena. The optimal conditions for studying motivational systems are therefore those in which the animal is in a familiar and unchanging external environment, and in relation to which learning is asymptotic.

7.2.1 Motivational State

The familiar though vague concept of motivational state can be given rigorous meaning on the basis of the usual definition of state variables (section 3.2.1 and section 6.1.1). The state of a system can be defined in terms of the state variables of the system. In Lorenz's model, for example, the state variables are the displacement X of the valve mechanism, and the amount of liquid in the reservoir, to which the height H is directly proportional.

Lorenz's (1950) model has been criticized on the grounds that the action-specific energy, the liquid in the reservoir, is "used up" in the performance of the behaviour pattern, whereas most of the evidence suggests that behaviour patterns are terminated as a result of various types of sensory feedback (Hinde, 1970). For example, Sevenster-Bol (1962) showed that the act of fertilizing the eggs is not necessary for the reduction of sexual behaviour in male sticklebacks, and that stimuli from the eggs alone are adequate. Feedback, however, is implicit in Lorenz's model as can be seen from the block diagram representation of the model, illustrated in Fig. 7.3. Failure to recognize this implicit feedback is probably responsible for the unnecessary criticism of Lorenz's concept of the "exhaustion" of a behaviour pattern when the reservoir is empty (see also section 7.1.2). Lorenz's contention, that the performance of the consummatory act *per se* is responsible for the exhaustion phenomenon, does not stand up to modern evidence. But in both Lorenz's formulation, and in the more overt feedback models, various consequences of the behaviour are responsible for the diminution of the motivational state responsible for the behaviour.

The motivational state of an animal should not be thought of in terms of unitary variables such as "drive level" (Hinde, 1959). There are likely to be a number of variables relevant to the state of each motivational system. In the case of the thirst system portrayed in Fig. 3.21 the animal receives information about the state of the blood and about the gut contents. It is reasonable to infer that both these, and other, state variables contribute to the thirst "drive" of the animal. Thus it should be possible to portray motivational state in an n dimensional state space (section 6.1.3), as shown in Fig. 6.8.

In general, motivational state variables can be divided into two classes, relating to the primary storage aspect of motivation, and to those aspects concerned with ongoing behaviour. The concept of storage is implicit in most theories of motivation. In Lorenzian theory it is equivalent to the accumulation of liquid in the reservoir, and is manifest in the phenomenon of "damming up" (e.g. Sevenster, 1961). In Hullian theory (see Bolles, 1967) storage is implicit in the direct relationship between drive D and the animal's physiological needs. Thus during food or water deprivation certain physiological variables become progressively displaced from their normal values and the drive accordingly increases.

McFarland (1970c) distinguishes between storage effects associated with the potentiality of the animal to perform certain types of behaviour and those associated with ongoing behaviour. The change in motivational state induced in an animal deprived of appropriate external stimuli may be said to lead to an increase in the animal's potential to carry out the behaviour, and when the appropriate external stimuli are presented, the intensity of the behaviour is generally an increasing function of deprivation. On the other hand, an animal prevented from continuing with its ongoing behaviour, in the presence of the external stimuli, is generally said to be in a state of frustration, rather than deprivation (Yates, 1962). In such cases an increase in the intensity of the behaviour can be observed when the obstruction is removed, a phenomenon generally known as the "frustration effect" (Amsel, 1958).

The distinction between these two aspects of motivation can also be made on other grounds, perhaps best illustrated by reference to the homeostatic type of motivational system, such as thirst. A general picture of a homeostatic motivational system is given in Fig. 7.18. Physiological imbalances occur both as a result of the action of environmental factors, such as temperature; and as a result of influences from other motivational systems, such as the feeding system. These imbalances are monitored by the central nervous mechanisms, which in turn actuate two types of corrective

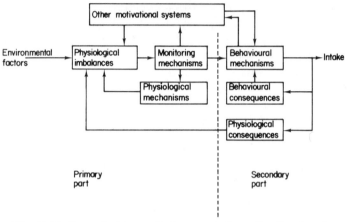

Fig. 7.18. General picture of a homeostatic motivational system.

mechanism. The latter may be conveniently classified into physiological and behavioural mechanisms which essentially act in parallel to correct the imbalance. An example of a physiological corrective mechanism would be the pituitary-kidney antidiuretic axis. Mechanisms of this type act to conserve the commodity in imbalance but are not always able to restore the balance. This is the prime function of the behavioural mechanism, the action of which results in intake of the required commodity. Such intake can have three types of effect: (1) it can have purely behavioural consequences which feed back to the behavioural mechanism, and serve satiation, reward, etc.; (2) it can have physiological consequences which act to restore the balance; (3) it can influence other motivational systems. For example, ingestion of cold water can have thermoregulatory consequences. This outline of a motivational system is brief and incomplete, but more detailed accounts of the processes mentioned here may be found in section 3.3 and section 6.3.2. For the present purposes a general outline is sufficient.

The motivational system portrayed in Fig. 7.18 falls into two distinct parts, separated by a dotted line. The primary part is continuously active and its state at any time represents a situation that can be called the "primary motivational state". The secondary part is active only when the animal is engaged in the appropriate type of behaviour. The "secondary motivational state" refers to those aspects of motivation involved in ongoing behaviour. Interactions between motivational systems can exert their effects on either the primary or the secondary parts. The state of the feeding system can affect the degree of build-up of motivational potential for drinking, but feeding behaviour itself can also affect drinking directly by its action on the secondary part. Thus it is useful to make a distinction between primary drinking and secondary drinking (see Fitzsimons, 1968). The primary and secondary aspects of motivation can also be differentiated by the methods employed in their study. Generally, the procedure employed in studies of the primary part is to take food and/or water deprivation time as an independent variable and to keep physiological and behavioural measures as standardized as possible. Conversely, the secondary part is studied by means of standardized deprivation schedules, whilst manipulating variables directly related to ongoing behaviour during recovery from deprivation. The methods employed and results obtained in experiments of this type have been reviewed extensively by Bolles (1967).

Quantitative estimation of motivational variables is a necessary precursor to formulation of a motivational system in terms of control theory. McFarland (1965c) obtained similar deficit functions in doves from measurements of body weight and food intake during water deprivation and water intake during recovery from deprivation. These functions provided essential evidence for the initial analysis of the dove drinking system in terms of control theory. On the basis of this type of analysis McFarland (1970c) has argued that it is possible to describe motivational systems in terms of the generalized variables and parameters discussed in section 1.1. The main arguments used to justify this type of formulation are that the existence of dynamical analogies between physical systems enables us to define generalized variables and parameters, in terms of any dynamic system, including a behavioural one. Description of motivational systems in generalized terms can be achieved through the application of control theory to the analysis of behaviour (see McFarland, 1970c for further details).

A generalized formulation of a simplified homeostatic system is illustrated in Fig. 7.19. This is similar in principle to Fig. 7.18, with the exception that interactions with other motivational systems have

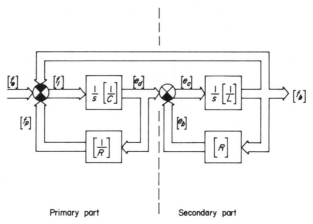

Primary part | Secondary part

Fig. 7.19. Generalized homeostatic motivational system. $[f_e]$ = environmental flow vector, $[f_p]$ = physiological flow vector, $[f_b]$ = behavioural flow vector, $[f_l]$ = net (loss) flow vector, $[e_d]$ = deficit effort vector, $[e_c]$ = command effort vector, $[e_b]$ = behavioural effort vector. Capacitance $[1/C]$, inductance $[1/L]$, and resistance $[1/R]$ matrices represent the respective parameters of the system

been omitted for simplicity. The physiological imbalances of Fig. 7.18 can be represented by the matrix equation

$$[x] = \frac{1}{s}[f_l] = \frac{1}{s}\Big([f_e] - [f_b] - [f_p]\Big) \tag{7.3}$$

where the vector $[x]$ represents the set of systemic physiological displacements, relevant to the homeostatic system in question. Each x is a state variable, analogous to Hull's (1943) need states, and results from integration of the net rate of loss of the relevant commodity, represented as the generalized flow variable f_l. The net rate of loss is simply the difference between the rate of loss due to environmental factors $[f_e]$, which include factors from other motivational systems, and the rates of gain resulting from intake $[f_b]$ and physiological conservations $[f_p]$. These are all generalized flow variables, analogous to mechanical velocity and electrical current (see Table 1.1).

A displacement cannot, of itself, act as a causal agent in a system. The causal agent, a rate variable, must be obtained by measurement of the displacement. For instance, the state of hydration of the blood does not, of itself, lead to drinking. It is measured by sensory cells which transmit information to the actuating mechanisms of the CNS. As the displacements *per se* are not actuating variables they are omitted from Fig. 7.19, and the integrating and measuring mechanisms are combined in the generalized capacitance, the output of which is the generalized effort e_d (see Fig. 6.6).

$$[e_d](s) = \frac{1}{s} \left[\frac{1}{C}\right]\left[f_l\right]$$ (7.4)

The outputs from the monitoring mechanisms (Fig. 7.18) can thus be seen as exerting an effort on the behavioural mechanisms responsible for ingestion, and on the physiological mechanisms responsible for conservation and storage. The physiological mechanisms can be represented as generalized resistances $[1/R_p]$, which have flow variable outputs $[f_p]$ representing the physiological consequences of the instructions to the CNS. In the case of the pituitary-kidney antidiuretic axis, for example, the applied effort has the physiological consequence that water is reabsorbed from the kidney tubules, thus decreasing the net loss.

The distinction between physiological displacements and a corresponding neurological variable, which is identified with the concept of effort, is similar to Hull's (1943) distinction between physiological needs and drive stimuli. But Hull's drive stimuli have no effect upon motivation *per se*. They give direction to motivated behaviour by their ability to acquire habit strength. Motivation is provided by the drive D which is reponsible for the energizing aspect of motivation and has no stimulus properties. The distinction between directing and energizing aspects of motivation is an unnecessary one, brought about by a misunderstanding of the energy concept, as we shall see in section 7.2.2.

The effort variables of Fig. 7.19 are equivalent to what some psychologists call drive, in that they actuate the mechanisms responsible for behaviour. It should be noted that at least three types of effort can be distinguished, each being a vector quantity. This formulation of the drive concept is in no way unitary. The effort vector $[e_c]$ responsible for actuating the behavioural mechanisms of Fig. 7.18 is distinguished from that $[e_d]$ resulting from the measurement of physiological state, by the intervention of the effort vector $[e_b]$ resulting from the measured consequence of behaviour. At this point the diagram (Fig. 7.18) is oversimplified, because some effort variables will have positive, and others negative, effects (see section 3.3.3), and influences from other motivational systems also operate at this point (see section 7.2.3). The behavioural mechanism, having an effort vector input, and a flow vector output, and having capacity for storage (see above), is identified with a generalized inductance (see Fig. 6.6). Integration of a generalized effort variable gives the generalized momentum p, analogous to mechanical momentum, and electrical flux linkage (see Table 1.1).

$$[p] = \frac{1}{s} [e_c] \quad \text{and} \quad [f_b](s) = \frac{1}{s}\left[\frac{1}{L}\right] [e_c] \tag{7.5}$$

Specific examples, involving formulation of motivational systems in terms of generalized variables and parameters are given in some detail by McFarland (1970c), and some consequences of this type of formulation are discussed later in this chapter. For the present, it is worth noting that there are two types of state variable by which the state of a motivational system may be characterized: the generalized displacement x, and the generalized momentum p.

The generalized displacement x characterizes the state of the physiological variables which are directly relevant to motivation. They are generally those aspects of the physiological background that are monitored by the CNS. In the case of homeostatic systems, such as hunger and thirst, the situation is fairly straightforward, but with sex, exploration and other non-homeostatic systems, a number of problems arise. One problem is whether the physiological antecedent conditions for sex and exploration are comparable with those of hunger and thirst. This question is reviewed at some length by Bolles (1967). In the case of sex, the hormonal state of the animal can be readily compared with its state of hunger or thirst, but in the case of exploration, fear, etc., the evidence is controversial (Bolles, 1967; Hinde, 1970), and it may well be that we cannot speak of primary motivational states for such behaviour.

Hormonal state influences aggressive, sexual, parental, and other types of behaviour (Hinde, 1970). The factors involved are extremely complex and the reader should refer to the relevant literature for details (e.g. Lehrman, 1961; Young, 1961; Hinde, 1970). In general, the systemic hormonal level is directly or indirectly under CNS control (Brown-Grant, 1966) and some progress has been made in analysing the mechanisms involved in terms of control theory (Yates et al., 1968; Gann et al., 1968). Physiological control of hormonal state is little different in principle from that of hunger or thirst, and a possible scheme is illustrated in Fig. 7.20. The systemic level of the many hormones involved can be represented by a state vector $[x_h]$ which results from the integration of the net rates of change of hormonal level $[f_n]$. Factors influencing the net rates of change include endogenous metabolic factors $[f_e]$, regulatory influences exercised by the CNS $[f_p]$ and influences of behavioural origin $[f_b]$. The systemic hormonal state $[x_h]$ monitored by the CNS, produces the effort vector $[e_d]$ which in turn affects the behavioural and physiological control mechanisms. We can see how the control of various types of hormonally influenced behaviour might be described in terms of generalized variables and parameters. The demonstration that such a formulation is possible must await an analysis of the behaviour in terms of control theory.

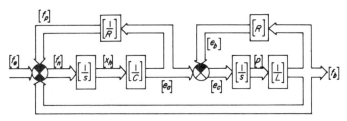

Fig. 7.20. Generalized hormonal motivational system. $[x_h]$ = hormonal displacement vector, $[p]$ = behavioural momentum vector. Other symbols as in Fig. 7.19.

The generalized momentum p characterizes the state of that part of the motivational system concerned with ongoing behaviour. As the behaviour relevant to any one motivational system is likely to be composed of a number of different activities, the momentum is best formulated as a state vector $[p]$.

The notion of momentum as a behaviourally relevant variable can be developed in a number of ways. Firstly, it follows logically from the fact that a storage phenomenon exists which is different from that associated with primary motivational state. Evidence for such a phenomenon comes primarily from work on frustration and related phenomena such as behavioural contrast (Bloomfield, 1968). The increase in intensity of ongoing behaviour following frustration has been demonstrated in a number of situations and this frustration effect has been attributed to generalized drive, incentive, and other theoretical constructs (Bolles, 1967). McFarland (1966c, 1970c) suggests that the frustration effect is an example of the "try harder" phenomenon exhibited by many error-actuated feedback systems in response to disturbance of the output. McFarland argued that ongoing behaviour is likely to be partly controlled by certain consequences of that behaviour which are monitored by the animal, as shown in Fig. 6.2. A simple feedback mechanism of this type would allow the animal to maintain its behaviour at a "desired" level, despite disturbances in the environment. Evidence for this type of system comes from Chung's (1965) study of the effects of effort on response rate in a Skinner box, and analysis in terms of control theory (McFarland, 1966c) provides a basis for the formulation of the mechanisms controlling ongoing behaviour in terms of generalized variables and parameters (McFarland, 1970c).

The necessity for a momentum concept can also be argued from consideration of the fact that animals are always motivated to carry out a number of incompatible behavioural activities. Once an animal has started on one activity, feedback from that activity tending to reduce the level of motivation and fluctuations in the level of motivation for other activities, will tend to disrupt the ongoing behaviour because of motivational competition (see section 7.2.3). Consequently, the animal will be continually "dithering", unable to complete any task. To prevent dithering it is necessary that the ongoing behaviour gain some momentum so that, once an animal has started on a particular course of action, it will continue. The necessary momentum could be provided by positive feedback from consequences of behaviour. For example, positive feedback from oral factors involved in feeding and drinking probably serves such a function (see section 3.3.3).

Direct observation of behaviour can also give clues to the operation of behavioural momentum and in this respect Lorenz's (1950) observations are interesting:

"When we suddenly deprive an animal of the object of its reaction, the activity never breaks off abruptly but nearly always continues a considerable time *in vacuo*. Doubtless it is a consequence of the same phenomenon that the 'momentum' gained by an activity will carry it on for an appreciable time after the moment when its releasing threshold, rising continually throughout the duration of the discharge, has reached the value corresponding to the external stimulation impinging at the moment."

Notice that Lorenz realized that the displacement of the spring (Fig. 7.1) gradually shifted as the head of liquid in the reservoir was lowered.

"The inertia of a reaction carries the threshold high above the value corresponding to the external stimulation impinging at the moment."

Lorenz goes on to argue that the system is underdamped, so that rhythmic activity can occur in the presence of constant external stimulation (see Fig. 7.3). Lorenz clearly identifies the phenomenon of inertia with the properties of the releasing mechanism. However

"It is probable that the 'initial' inertia of endogenous activities and their propensity to continue longer than corresponds to present stimulation, are two entirely different phenomena, requiring different physiological explanations. The way in which quiescent activities respond to stimulation is more suggestive of initial 'friction'. They 'behave' exactly as if the valve releasing their discharge were a bit sticky. Furthermore, there arises the question whether there is not a distinct relationship between creeping in of stimulation and what we call habituation. Contrary to these phenomena of an initial resistance to stimulation, the continuance of an activity after cessation of adequate stimulation is easily explainable on the assumption that the activity is 'self-stimulating'."

7.2.2 Motivational Energy

Most theories of motivation have involved an energy concept in one form or another. Hinde (1960) reviews the energy models of motivation used by Freud (1940), McDougall (1923), Lorenz (1950) and Tinbergen (1951), enumerates the confusions and inconsistencies involved in such models and doubts whether an energy concept is necessary. Bolles (1967) claims that, as behaviour becomes better understood, there emerges a variety of ways of accounting for what appears to be an energizing effect of motivation without invoking an energization principle. On the other hand, McFarland (1970c) argues that a rigorous definition of motivational state makes an energy concept logically inevitable, although whether such a concept will prove to be a useful one is an open question. McFarland's argument may be summarized as follows: (1) the existence of dynamic analogies between physical systems enables us to define generalized variables and parameters, in terms of which any dynamic system may be described. (2) Description of motivational systems in generalized terms can be achieved through the application of control theory to the analysis of behaviour as a function of time. (3) By delineating the important state variables in a motivational system the concept of motivational state can be given rigorous meaning. (4) Consideration of the changes of state involved in behavioural transients leads inevitably to concepts of work, power and energy, analogous to those in the physical sciences.

System variables can be divided into state variables and rate variables (see section 3.2.1). Rate variables are involved in the process of changing the state of a system, the measure of which is defined as "work". "Power" is defined as the rate of doing work and is the product of two rate variables. Energy, defined as the capacity for doing work, is the time integral of power, viz., $P = dE/dt$. Flow is usually opposed by a resistance, which is overcome by applying a suitable effort. In this process energy is dissipated or lost from the system to its environment. Thus the generalized resistance is defined in terms of the generalized rate variables, $R = e/f$, and the rate of energy dissipation through the resistance is the power $P = ef$. The generalized energy storage elements are the generalized "capacitance" C, measured in terms of displacement per unit effort

$$\left(\text{thus } x = Ce \quad \text{and} \quad e = \frac{1}{C}\int fdt\right);$$

and the generalized "inductance" L with units of momentum per unit flow

$$\left(\text{i.e.} \quad p = Lf \quad \text{and} \quad f = \frac{1}{L}\int edt\right).$$

The generalized capacitance stores "potential" energy when its effort level is raised by influx of a flow variable. Thus we can write displacement

$$x = \int fdt,$$

potential energy

$$E_p = \int_0^x edx = x^2/2C = \tfrac{1}{2}Ce^2 . \tag{7.6}$$

Similarly, the generalized inductance stores "kinetic" energy in accordance with the following equation, momentum

$$p = \int edt,$$

$$E_k = \int_0^p fdp = p^2/2L = \tfrac{1}{2}Lf^2 . \tag{7.7}$$

In terms of the generalized systems illustrated in Figs 7.19 and 7.20, when the primary and secondary parts are separated, as during deprivation, there will be changes in the systemic physiological displacements [x] as a result of the action of motivational flow variables (eqn. 7.3). These changes may not be independent of each other. For instance, a displacement relating to the blood glucose level may be dependent upon, or constrained by, a displacement relating to systemic osmotic pressure. The details of the constraints operating at this level will be obviously dependent upon the particulars of the physiology of the animal. However, it is important in considering energy functions that the coordinates of the state space, in this case the primary motivational state space, be independent and completely unconstrained. A variety of coordinates may be employed in considering the state of a system, and as a matter of convenience the letter q is employed as a symbol for coordinates regardlesss of their nature. Thus q is referred to as a "generalized coordinate" (Wells,

1967). The constraints operating in a system can often be expressed in terms of integrable differential or algebraic equations in which case they are known as "holonomic" constraints. Non-integrable equations of constraint characterize non-holonomic constraints. Suppose that a system has N_u state variables, and there are N_c holonomic constraints, then the number of generalized coordinates N_q required to describe the state of the system is given by the difference $N_u - N_c = N_q$, and this number is known as the "degrees of freedom" of the system (Shultz and Melsa, 1967). In a (holonomic) motivational system therefore the generalized coordinates q are those displacements x, between which there are no constraint relations.

Returning to the generalized system illustrated in Fig. 7.19, when the primary and secondary parts are separated, the effort level $[e_d]$ of the system is raised by influx of environmental flow variables, in accordance with eqn. 7.4. This change in effort level implies corresponding changes in physiological displacements $[x]$ and a change in energy level analogous to that described by eqn. 7.6. During water deprivation, for instance, the level of motivational energy is raised as a result of continued dehydration $[f_1]$. Because the energy storage elements are identified with generalized capacitances, the stored energy is analogous to the "potential energy" of a physical system (Table 1.1). As a rule, a motivational system will store potential energy when an actuating variable is displaced from its normal (steady-state) value. As there are likely to be a number of relevant actuating variables in any motivational system the total potential energy will be made up of many terms, of which each may be a function of more than one variable. In general

$$E_p([q]) = \int_0^q \sum_{i=1}^{N} e_i([q])dq \qquad (7.8)$$

where $E_p([q])$ is the total potential energy of the motivational system, and N is the number of generalized coordinates. Although potential energy is denoted as a function of the vector $[q]$, this does not imply that it is a vector quantity but that the potential energy is to be evaluated as a line integral from the origin to an arbitrary point $[q]$ in the generalized state space. When $N = 3$ the line integral may be evaluated successively in three dimensions of the state space (Fig. 7.21). Thus eqn. 7.8 may be rewritten in the expanded form,

$$E_p([q]) = \int_0^{q_1} e_1(q_1, 0, 0)dq_1 + \int_0^{q_2} e_2(q_1, q_2, 0)dq_2 +$$
$$\int_0^{q_3} e_3(q_1, q_2, q_3)dq_3. \qquad (7.9)$$

This simple method of evaluating the line integral is possible because $E_p([q])$ is a state function and the line integral is therefore independent of the path of integration (Shultz and Melsa, 1967).

If the action of the environment were uninhibited during deprivation (i.e. $[f_i] = [f_e]$ in Fig. 7.19), then all the input would be stored as motivational energy. Yet inspection of Fig. 7.19 shows that $[f_i] = [f_e] - [f_p]$ during deprivation. The action of the physiological loop thus reduces $[f_i]$ in proportion to the level of $[e_d]$ and the amount of energy stored is consequently reduced by the action of this feedback path. In generalized terms, energy is dissipated through the resistance matrix $[1/R_p]$, the total energy loss being the sum of the losses in each of the elements of the matrix. For a particular resistance element $1/R_n$ the energy loss

$$E_n(t) = \int_0^t (f_p e_d)dt = \int_0^t P_n dt \qquad (7.10)$$

where $P_n = f_p e_d$ is the instantaneous power loss associated with the resistance element. It is possible, then, to think of motivational energy being dissipated through the action of physiological processes subserving a conservation function and some implications of this concept are discussed by McFarland (1970c).

From a long-term viewpoint, the behavioural element in a motivational system can be regarded as a zero-order sub-system (see section 3.3.1). This could be represented in general terms as a resistance matrix $[1/R_b]$ through which potential energy is dissipated during ongoing behaviour, in a manner similar to that outlined in eqn. 7.10, for the physiological resistance matrix. In recovery from deprivation the behavioural sub-system is effectively the recipient of a step input from the primary part of the motivational system (see section 3.3.1) and is thus forced to change its state. The level of motivational energy of the behaviour mechanism is raised as the result of this state change, just as the level of potential energy is raised following a change in primary motivational state (eqn. 7.8). Because the energy storage element is identified with a generalized inductance (Fig. 7.19), the stored energy is analogous to the "kinetic energy" of a physical system (Table 1.1). This analogy enables us to formulate inertial forces $d/dt[f][L]$, involved in the initiation of a behaviour pattern, and of that behaviour developing momenta $[p]$. However, having defined the generalized coordinates q, it is convenient to determine the kinetic energy in the same terms, rather than in terms of the state variable p, as in eqn. 7.7. Thus

$$E_k([q], [\dot{q}]) = \int_{0}^{q} \sum_{i=1}^{N} p_i([q], [\dot{q}])d\dot{q}_i \qquad (7.11)$$

where $\dot{q} = dq/dt$, $E_k([q], [\dot{q}])$ is the total kinetic energy of the motivational system, and N is the number of generalized coordinates. The behaviour of the animal can thus be expressed as a trajectory in a generalized flow $(\dot{q})$ space and the kinetic energy evaluated by

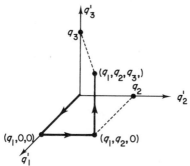

Fig. 7.21. Simple path of integration in a three-dimensional q space. (From Schultz and Melsa, 1967.)

taking a line integral corresponding to that illustrated in Fig. 7.21. When, for example, $N = 3$

$$E_k([q], [\dot{q}]) = \int_{0}^{\dot{q}_1} p_1([q], \dot{q}_1, 0, 0,)d\dot{q}_1 +$$

$$\int_{0}^{\dot{q}_2} p_2([q], \dot{q}_1, \dot{q}_2, 0)d\dot{q}_2 + \int_{0}^{\dot{q}_3} p_3([q], \dot{q}_1, \dot{q}_2, \dot{q}_3)d\dot{q}_3 \qquad (7.12)$$

In terms of Fig. 7.19, the slow changing $[f_e]$ and $[f_p]$ can be regarded as constant during the relatively short periods involved in recovery from deprivation. Thus in terms of generalized variables

$$[x] = \frac{1}{s}[f_1] \triangleq \frac{1}{s}[f_b], \qquad (7.13)$$

and in terms of generalized coordinates

$$[q] = \frac{1}{s}[\dot{q}] \qquad (7.14)$$

Thus the kinetic energy is evaluated in terms of those flow variables $[f_b]$, which are directly related to the primary motivational state variables. A number of problems of detail remain to be worked out, particularly concerning the relation between kinetic energy and short-term consequences of behaviour. The present aim is to introduce the general principles involved in the concept of motivational energy and if the kinetic energy concept seems more vague than that of potential energy it should be remembered that motivational aspects of ongoing behaviour have long been neglected, compared with the study of primary aspects of motivation. It is here envisaged that kinetic energy storage is associated with ongoing behaviour and its immediate consequences. There will be an alteration in kinetic energy storage whenever the consequences of ongoing behaviour are changed. Thus the "frustration effect" might be seen as a kinetic energy phenomenon, as might any behavioural phenomenon which relates to the rate at which behaviour consequences affect the primary motivational state.

The main problem posed by the concept of motivational energy is the behavioural meaning of energy dissipation. Energy is dissipated through a behavioural resistance, when the energy level of the system is reduced. The problem is aggravated by the fact that the so-called behavioural resistances are in fact sub-systems containing smaller energy storage elements, the behaviour of which demands more detailed analysis. In general we can say that, as in physical systems, the kinetic energy level of a sub-system is increased temporarily as potential energy is dissipated through it. But, as in physical systems, the usefulness of such concepts can be evaluated only through rigorous quantitative experimental analysis of the behaviour of the system. Although some steps have been taken in this direction (McFarland, 1970c), validation of the formulation outlined here must await further empirical verification.

At this point it may be useful to look in greater detail at some of the criticisms levelled at earlier formulations of motivational energy. The principal critique is that of Hinde (1960). Hinde points out that the energy concepts of McDougall (1923), Freud (1932, 1940), Lorenz (1937, 1950) and Tinbergen (1951) differ in many ways, but all share the idea of a substance capable of energizing behaviour, held back in a container and subsequently released in action. Hinde's criticisms are concerned with: (1) the role of energy as a substance, which "flows", "accumulates", etc. (2) the physiological embodiment of the energy concept and (3) confusion between physical energy and its behavioural analogy.

Lorenz (1950) provides the prime example of the concept of energy as a substance. He refers to the liquid in the reservoir as "action-specific energy", but he attributes no properties to this postulated entity, which could not also be attributed to a material substance. Lorenz's model is an energy model in name only. According to Hinde (1960), the common view that cessation of activity is a consequence of the discharge of energy, arises from the fact that the properties of physical energy are imputed to behavioural energy. This is not necessarily so, since Lorenz need not have used the word "energy" at all, but would still have attributed cessation of activity to the discharge of a substance from the reservoir. However, the substantive criticism on this issue is that energy dissipation is incompatible with the view that cessation of many activities is due to the action of consummatory stimuli, which feed back and inhibit the response (Hinde, 1970). The view that energy models leave no room for inhibition (Kennedy, 1954) is valid only if energy is conceived of as a substance. In the energy formulation outlined above (eqns. 7.6-7.12) feedback from the consequences of behaviour is necessary for behavioural energy dissipation, rather than the reverse. What, therefore, is the main difference between this and previous types of energy formulation, that makes this kind of criticism invalid? Most energy models of motivation employ energy as a causal agent, which stops, starts or controls behaviour. No such causality is implied by the present energy concept. Indeed, the energy formulations result from an understanding of the causal factors responsible for the behaviour, rather than the reverse.

Hinde's second type of criticism is implicit in his discussion of the extent to which the energy models are considered by their authors to correspond to structures in the nervous system. Lorenz's (1950) model is based on behavioural data, and physiological evidence is used only *post hoc.* Freud's (1940) model developed from physiological considerations, but subsequent Freudian theory bears little or no relation to physiology. Tinbergen (1951) refers to his hierarchical system of Lorenzian reservoirs as "a graphic picture of the nervous mechanisms involved", and clearly regards his "centres" as neural structures. McDougall (1923) also regards the relationship between energy and the nervous system as a close one. He envisaged that energy liberated on the afferent side of the nervous system was held back by "sluice gates" in the optic thalamus. Such confusions between "hardware dependent" and "hardware independent" explanations are common, not only to energy models, but also to other types of cybernetic model. The distinction is discussed in the preface to this book.

Hinde's third type of criticism is that the concept of behavioural energy has in many cases become confused with that of physical energy. Lorenz's (1950) model is free from this infection, retaining its independent analogical status, but Tinbergen's (1951) model is clearly contaminated. Thus Tinbergen writes of thwarted energy "sparking over" to another behavioural system, when its discharge is blocked. Freud (1940), recognizing the danger, wrote of mental energy, "We have no data which enable us to come nearer to a knowledge of it by analogy with other forms of energy," but subsequent psychoanalysts (e.g. Colby, 1955) postulate mental energy as a form of physical energy. McDougall (1913, 1923) also regarded "psycho-physical energy" as a form of physical energy. Hinde (1960) discusses the way in which the properties of the model come to be confused with those of the original (see also Meehl and McCorquodale, 1948): "Such properties are introduced surreptitiously as occasion demands, and involve a transition from admissible intervening variables, which carry no existence postulates, to hypothetical constructs which require the existence of highly improbable entities and processes." A behavioural energy concept is admissible as an analogy only and "the question of convertibility into physical energy is a dangerous red herring". In the case of the present formulation of motivational energy, which is strictly analogous to that of the physical sciences, the question of convertibility does not arise. But is the issue a red herring in another sense? The mathematical validity of the motivational energy concept does not necessarily mean that it is a useful one.

The energy concept has considerable intuitive appeal, as evidenced by its frequent use in the past. Such concepts as energy storage, and behavioural inertia and momentum which make behavioural sense intuitively, need not be used vaguely and can be given a precise meaning in terms of the observed and postulated variables involved. By introducing a motivational energy concept, the way is opened for *post hoc* identification of the energy exchanges involved. For example, the vacillations observed in conflict situations immediately suggest a reciprocal exchange of potential and kinetic energy, and even if it is possible to analyse the system in a conventional manner, it may be possible to work backwards from this supposition. Moreover, energy concepts are becoming increasingly important in control theory in general. The description of behavioural systems in terms that are relevant to control theory may open up the way to a general method of behaviour analysis based upon direct observation of behaviour and this approach is discussed further in section 7.3.3.

7.2.3 Motivational Interactions

Motivational interactions have posed problems for many theorists. Hull's (1943) distinction between energizing and directing aspects of motivation is a prime example. The energizing aspect depends upon the strength of the generalized drive D, to which all the specific need states contribute and which is capable of activating all acquired habits in the presence of the appropriate external stimuli, regardless of the specific need state under which the habits were acquired. The directing aspect of motivation is due to the drive stimuli, which arise from the physiological needs, but have no effect upon motivation *per se.* They give direction to motivated behaviour by their ability to acquire habit strength. For Hull, drive interaction presents no problem, because all drives summate. The validity of the general drive concept has frequently been questioned on empirical grounds (Bolles, 1958; Zeigler, 1964; McFarland, 1966b). After reviewing the evidence, Bolles (1967) concludes that "behaviour is predominantly determined by the specific drive conditions, specific stimulus situations, and specific habit structures that characterise an individual at any given time", and that the general drive factor is only a minor determinant of behaviour. Similarly, Hinde (1970) concludes that the intensity of any one type of behaviour depends only on motivational factors more or less specific to it and that general factors are of minor importance.

While the concept of general drive has been widespread amongst psychologists, ethologists have tended to assume that the total behaviour repertoire of an animal is made up of a number of distinct motivational systems, which are specific to their respective categories of behaviour. Lorenz's (1950) model is intended to represent such a specific motivational system and one difficulty with the model is that it is difficult to envisage interactions between such systems. Tinbergen (1951) attempted to do this by arranging Lorenzian systems in a hierarchical fashion, but this manoeuvre gets him into analogical difficulties (Hinde, 1956, 1960). Zeigler (1964) points out that the concept of general drive is implicit in Tinbergen's "sparking over" explanation of displacement activities in which motivational energy is diverted from one behavioural system to another. However, there is an important difference between Tinbergen's formulation and general drive theory. According to the latter, one type of drive can at all times activate other types of behaviour, while according to Tinbergen's theory this can only take place when the ongoing behaviour is blocked (McFarland, 1966b).

Much of the recent literature on drive interaction is reviewed by Campbell and Misanin (1969). These authors make no distinction between the different levels at which such interaction can take place, and their account is consequently rather confused. Von Holst and von Saint Paul (1963) postulated various levels of interaction, based upon behaviour elicited by various intensities of electrical stimulation of the brain of domestic fowl. In terms of the type of system illustrated in Fig. 7.19, three main levels of interaction may be distinguished. These may be called the primary level, the secondary level, and the level of the "final common path" (von Holst and von Saint Paul, 1963).

At the primary level, the environmental influences may come directly from the environment as instanced by the effect of ambient temperature upon the temperature regulation system, but more commonly they come from other motivational systems. Thus the effect of ambient temperature upon feeding and drinking operates via the thermo-regulatory system. The environmental flow vectors can be divided into three main types: (1) those originating from the environment proper [f_e]. So rate of heat loss is directly influenced by environmental temperature. (2) Those representing behavioural flow variables [f_b] from other motivational systems. For example, increased water loss can be a direct consequence of food intake (McFarland and Wright, 1969). (3) Those [f_p] originating as effort vectors [e_c] in other motivational systems, and operating via a physiological resistance matrix [$1/R_p$]. Thus particular motivational states can affect the release of hormones relevant to other systems. In some birds, for instance, incubation facilitates the production of prolactin which provides the hormonal background necessary for parental behaviour (Lehrman, 1961). Each of these possible types of effect is illustrated in Fig. 7.22.

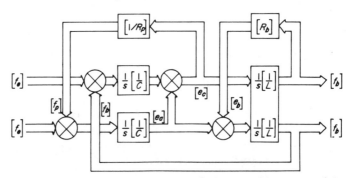

Fig. 7.22. Generalized representation of the possible types of interaction between motivational systems. Symbols as Fig. 7.19.

At the secondary level there are two main types of interaction: (1) behavioural flow variables $[f_b]$ acting through a behavioural resistance matrix $[R_b]$ may set up effort vectors $[e_b]$ relevant to other motivational systems. For instance, water ingestion may lead to changes in brain temperature (McFarland and Budgell, 1970a; Budgell and McFarland, 1971), and consequences of reproductive behaviour may lead to shifts in motivation by providing relevant external stimuli. Thus the condition of the nest and the presence of eggs in the nest contribute to shifts in motivational state (Lehrman, 1959; Hinde and Steel, 1966). (2) There may be direct action of effort variables in one motivational system affecting those in another. Thus the evidence suggests that the depressive effect of thirst on feeding is due to inhibition within the CNS (McFarland, 1964; Oatley and Tonge, 1969). Interactions of this type may be shown to be independent of motivational competition at the level of the final common path. Epstein, Fitzsimons and Rolls (1970) observed that a starving rat, which has just been allowed to start eating, changed to vigorous drinking when injected with angiotensin intracranially. To demonstrate that angiotensin genuinely inhibits feeding, it is necessary to show that the effect will occur in the absence of external stimuli associated with drinking, attention to which could compete with those of feeding. This brings us to consideration of the third type of motivational interaction.

Behaviour can be classified into various mutually exclusive categories, such that behaviour belonging in one category is incompatible with that belonging in another (section 3.1.2). For example, an animal does not normally indulge in feeding and sexual behaviour at the same time. Feeding and sexual behaviour can be said to belong to different motivational systems.

At any particular time, many motivational systems will be active, and these must in some sense compete for the "behavioural final common path", so called because it involves the last type of interaction in the causal chain, the final decision before the potential activity becomes overt. This term is strictly relevant to consideration of motivational systems in behavioural terms rather than the structure of the nervous system. Von Holst and von Saint Paul (1963) used the term "initial common path" for competition at the perceptual level, and "final common path" (after Sherrington, 1906) for competition at the motor level. Both types of competition are envisaged as operating in the behavioural final common path of a motivational system (McFarland, 1966a, b) (see Fig. 7.22).

The level of causal factors for some systems will be greater than that of others, so that the animal may be said to have a set of "motivational priorities". Feeding, then, might be the top priority activity, grooming the second-in-priority, and sleep the third. We can expect the order of priority to be related to the sequence of overt activities that are observed in a particular situation, especially when there is little change in external stimulation. The question is: what factors determine which motivational system is to have priority in the behavioural final common path and thus gain overt expression?

The most simple answer to this question is that the system which has the highest level of causal factors gains priority by virtue of "competition" with the other systems. The other systems are then said to be subject to "behavioural inhibition". According to Hinde (1970),

"Behavioural inhibition is . . . said to occur when the causal factors otherwise adequate for the elicitation of two (or more) types of behaviour are present, and one of them is reduced in strength because of the presence of the causal factors for the other. . . . In practice, since it is usual for causal factors for more than one type of behaviour to be present, some degree of motivational inhibition probably occurs all the time."

The amount of time an animal spends at a particular activity may thus be restricted by the necessity for doing other things. For instance, the proportion of the day spent feeding by blue tits increases from about 70% in summer to 90% in winter and there is a correlated decrease in the time spent resting and preening (Gibb, 1954). Cotton (1953) studied the effect of food deprivation on running speed in rats trained to run for food reward. Overall running speed was found to decrease with increased deprivation time, but the effect was much less marked when the time spent at "competing responses", such as grooming and sniffing, was subtracted from the overall score. The most important effect of deprivation was the increased priority of running for food, and a similar effect was found by Cicala (1961).

Taken at face value, these findings suggest that the causal factors for a particular activity build up to a level sufficient to oust the ongoing behaviour by motivational competition. Continued performance of a behaviour pattern produces consequences which feed back and reduce the level of causal factors to a point where the causal factors relevant to another activity are sufficiently strong to take over. Each motivational system follows a kind of relaxation oscillation, each "drive" finding its own level in a "free market" (Logan, 1964). But far from being self-regulatory, some motivational systems are partly under the control of permissive factors from other systems. *Ad libitum* drinking, for instance, is directed primarily by food intake in rats (Fitzsimons and LeMagnen, 1969; Kissileff, 1969) and doves (McFarland, 1969a).

Animals rarely indulge in one type of activity exclusively for long periods of time. Feeding, courting, running in a maze, are generally interspersed by other activities, which interrupt the primary ongoing behaviour. Many psychologists call such activities "competing responses" on the assumption that they arise by motivational competition, as outlined above. However, *a change in behaviour due to competition can in practice be recognized when a change in the level of causal factors for a second-in-priority activity results in an alteration in the temporal position of the occurrence of that activity* (McFarland, 1969a). Unless such a test is applied, the term "competing response" has descriptive validity only. Falk (1969) prefers the term "adjunctive behaviour", as it implies only that the observed behaviour occurs as an adjunct to a particular situation without evaluating its causation. Competition is characterized by the level of causal factors for a particular activity being ultimately responsible for the removal of behavioural inhibition on that same activity. The alternative is that the inhibition is removed by other factors, so that the causal factors for a particular activity play no role in the removal of inhibition on that activity. This is the essence of "disinhibition". Behavioural disinhibition as a mechanism responsible for the occurrence of adjunctive behaviour was originally (e.g. van Iersel and Bol, 1958; Rowell, 1961; Sevenster, 1961) envisaged as a consequence of conflict between incompatible activities. At the equilibrium point in a conflict the inhibition which the conflicting activities would normally exert on other behaviour patterns is removed, allowing a third (displacement) activity to "show through" (Rowell, 1961). There is evidence that such displacement activities occur when the competing predominant behaviours are equally balanced (Rowell, 1961; Sevenster, 1961), identifying the equilibrium point as the permissive factor in the situation. However, it is evident that conflict is not necessary for disinhibition to occur. Disinhibited activities can be identified in thwarting (McFarland, 1966a), feeding (McFarland and L'Angellier, 1966) and drinking (McFarland, 1970b) situations, and probably occurs in any situation where the behaviour sequence is stereotyped (McFarland, 1969a).

As a general rule, *the time of occurrence of a disinhibited activity is independent of the level of causal factors relevant to that activity* (McFarland, 1969a). A simple experiment may serve to illustrate this point. Barbary doves are food deprived and allowed to work for food in a Skinner box. After 5 or 10 min the feeding is interrupted by a bout of grooming behaviour, and then feeding is resumed. To test whether the grooming is due to competition or to disinhibition the experiment is repeated, under the same conditions, except that each bird is placed in the Skinner box with a paper clip fastened to the primary feathers on each wing. The results are the same as on the previous occasion except that the amount and vigour of preening is enhanced. In other words, raising the level of causal factors for preening, by means of paper clips, does change the intensity of the preening behaviour, but does not alter its time of occurrence. In this situation preening is a disinhibited activity (McFarland, 1970b). A number of workers have shown that adjunctive behaviour can be influenced by manipulation of the relevant causal factors. Thus Rowell (1961) showed that the intensity of displacement grooming in chaffinch could be increased by treating the plumage with water, or dirtying the bill with a sticky substance. Similarly, Sevenster (1961) showed that displacement fanning in the three-spined stickleback could be facilitated by increasing the carbon dioxide concentration in the region of the nest, and McFarland (1965b) found that displacement feeding in doves was affected by the presence of grain and the degree of hunger. However, this type of evidence alone is not sufficient to show that disinhibition has occurred. It is also necessary to show that the temporal occurrence of the activity in question is not affected by such treatment. This type of evidence was used by Sevenster (1961) in concluding that sexual fanning in the stickleback is a disinhibited activity, and by McFarland (1970b), who showed that adjunctive behaviour in feeding and drinking situations is primarily disinhibited.

The mechanisms responsible for behavioural disinhibition are by no means understood and the disinhibition phenomenon poses considerable problems for theories of behaviour control. In particular, the fact that the frequency of occurrence of behaviour relevant to one system can come under the control of other motivational systems, means that motivational interactions at the level of the behavioural final common path are much more complex than might appear at first sight (see section 7.3.2).

Interactions between motivational systems are extremely involved, and their division into three types should only be regarded as a first step in the process of analysis. It is highly probable that interactions at more than three levels occur. From the point of view of control theory, we have to consider not only the motivational states set up within the system being considered, but also the relevant states in other systems.

7.3 DECISION MAKING

The study of decision making provides a meeting point for many disciplines: statistics, economics, philosophy, psychology and control engineering. In statistical decision theory there are two important classes of variables. One, the subject's evaluation of the relative attractiveness of alternative choices, is called the "utility". The other, called the (subjective) "probability", concerns the subject's evaluation of the consequences of making each choice. Thus a man may wish to make a decision between going to a restaurant or to a cinema, the utilities. He may also take into account the probability of the restaurant being expensive, or the film being poor. The key questions of statistical decision theory have been summarized as follows (Edwards and Tversky, 1967):

1. How do men make judgements of the utility or attractiveness of various things that might happen to them, and how can these utilities be measured?

2. How do men judge the probabilities of events that control what happens to them and how can these judgements of probability be measured?

3. How are judged probabilities changed by the arrival of new information?

4. How are probabilities and utilities combined to control decisions?

5. How should psychologists account for, or think about, the fact that the same man, put in the same situation twice, will often not make the same decision?

Reports of attempts to answer these questions make up a voluminous literature which is admirably reviewed by Edwards (1954, 1961). Little of this work is directly relevant to animal behaviour because it involves assumptions about the nature of man which few behaviourists would be happy to apply to animals. The notion of probability, in particular, involves the assumption that the subject makes complex estimates of possible future consequences of certain courses of action. It may well be that the ability of animals to anticipate future consequences is more limited than the assumptions allow. While the concept of subjective probability plays an important part in statistical decision theory, it is difficult to see how such a concept could be useful in animal work. The concept of utility is likely to be of more direct relevance but work with animals raises the possibility of manipulating utility directly which avoids the necessity for complex *post hoc* arguments. Nevertheless, it is likely that concepts similar to utility and probability will prove important in animal work, though they may not necessarily be formulated in the same manner (see section 7.3.3).

A less anthropomorphic approach to decision making can be found in modern control theory. When a controller is required to operate in relation to a specified performance criterion, there may be more than one possible course of action. The problem is to find the course of action which best approaches the criterion. This basic "optimal control" problem can be divided into a number of types (Prime, 1969).

(1) The "deterministic/optimal control problem" can be illustrated by reference to Fig. 7.23a. Given the dynamic relationship

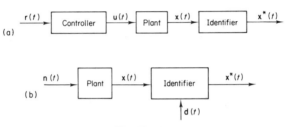

Fig. 7.23.

between $[x](t)$ and $[u](t)$, and between $[x^*](t)$ and $[x](t)$, and a performance criterion as a function of $[r](t)$ and $[x](t)$; the problem is to determine the control $[u](t)$ such that $[x](t)$, or $[x^*](t)$ fulfils the requirements of the performance criterion.

(2) The "estimation problem" can be illustrated by reference to Fig. 7.23b. Given the plant and identifier dynamics, the statistical properties of $[u](t)$ and $[d](t)$, the measurements of $[x^*](t)$ up to a certain time; the problem is to determine the best estimate of $[x](t)$, designated $[\hat{x}](t)$, in the sense of minimizing an objective function.

(3) The "stochastic/optimal control problem" can be illustrated by reference to Fig. 7.23a. Given problem (1) together with the statistical properties of disturbance vectors operating on the plant and identifier, determine the control $[u](t)$ such that $[\hat{x}](t)$, a best estimate of $[x](t)$, fulfils the specified performance criterion.

Three main methods are used in the solution of these optimal control problems. The first, based upon the use of calculus of variations is concerned primarily with the maxima and minima of arcs and surfaces in a multidimensional space. Variational methods are independent of the coordinate system used and can be applied in the determination of optimal trajectories in the state space of a system. A number of books on modern control theory contain an introduction to variational methods (Elgerd, 1967; Schultz and Melsa, 1967; Prime, 1969).

An alternative to the state vector representation of systems is a representation which makes use of energy and momentum concepts. Originally part of classical mechanics, the Lagrangian and Hamiltonian methods can be extended to systems analysis by generalization of the energy concept (see section 7.2.2). A development of this approach, the "Pontryagin maximum principle", has become an important part of modern control theory (see Schultz and Melsa, 1967; Prime, 1969).

A third approach to optimal control, originated by Bellman (1957), is known as "dynamic programming". Essentially a mathematical theory of multistage decision processes, it comes closer to standard statistical decision theory than do other types of optimal control theory. However, dynamic programming involves a series of attempts to move toward the optimal condition, each of which is primarily dependent upon the state of affairs resulting from the previous decision and not so much upon attempts to estimate future conditions. Dynamic programming is applied both to control problems (Prime, 1969) and to decision theory in general (Kaufmann, 1968) and is rapidly developing as a subject of study in its own right. In this respect, Jacobs (1967) provides a good introduction.

There has been very little application of decision theory, or optimal control theory, in studies of animal behaviour. Decision making is particularly pertinent to the study of motivation and there are some recent signs of activity in this neglected area.

7.3.1 Decision Theory and Motivation Theory

A major difference between decision theory and other types of cognitive theory is that decision theorists are concerned with the situation at the time, rather than in the past. Lewin (1936) represented the multiplicity of factors influencing behaviour as a psychological field. "One of the basic statements of psychological field theory can be formulated as follows: any behaviour or any other change in a psychological field depends only upon the psychological field *at that time*." (Lewin, 1943.) Lewin's distinction between historical and ahistorical explanations of behaviour is similar to that made in section 7.1.0 between learning systems and motivational systems. For Lewin, learning involves changes in the parameters of the psychological field especially in the "valence", or goal-attractiveness (Hilgard, 1956). Lewin's notion of valence is similar to the notion of utility in statistical decision theory (Edwards, 1954; Atkinson and Birch, 1970). Lewin *et al.* (1944) argued that the psychological force on an individual to undertake a certain behavioural activity is related to the subjective probability that the behaviour will lead to the goal, and to the valence of the goal. Lewin's ideas have been seen as a type of "expectance x value" formulation, similar to that of statistical decision theory (Atkinson, 1964; Atkinson and Cartwright, 1964).

A traditional premise of statistical decision theory is that the "expected utility" of an alternative is equal to the sum of the products of the subjective probability and utility of all the consequences of an act, and that the alternative having the greatest expected utility will be chosen from among a set of alternatives. This type of formulation has been incorporated into motivation theory by a number of workers particularly in the field of human motivation (Atkinson, 1964; Atkinson and Cartwright, 1964; Feather, 1963; Raynor, 1969). The most developed and generalized study is that by Atkinson and Birch (1970) on the "dynamics of action". These authors maintain that the "simplest change of activity in a constant environment implies a change in the dominance relations of tendencies during the interval of observation. This is the premise on which our theoretical analysis of the dynamics of action is based." (Atkinson and Birch, 1970.) These workers account for changes in the relative strength of tendencies in terms of three interrelated dynamic processes: "instigation", "resistance" and "consummation". The processes occur simultaneously to control the relative strength of the resultant action tendencies. Each observed activity is the expression of the strongest behavioural tendency at that time. The strength of this tendency is the algebraic resultant of the conflict between an action tendency to engage in the activity, and a negation tendency not to engage in the activity. Atkinson and Birch's conception of the instigation of action is a formulation of the way in which the strength of an action tendency is increased and sustained by the instigating force of the stimulus situation relevant to that activity. Their conception of resistance to action is a similar formulation of the effect of the stimulus situation on the negation tendency, which resists the behavioural expression of the action tendency, and is dissipated in so doing. Their conception of consummation relates to the functional significance of the behaviour. Consummatory factors reduce those action tendencies that are being expressed in behaviour and this function is referred to as the consummatory force of an activity. Atkinson and Birch's theory is complex, involving a large number of axioms and definitions. In its mathematical form it also involves a number of assumptions concerning linearity, additivity, etc. Particularly complex is the notion of resistance, which appears to invoke the basic concepts of decision theory (outlined above) in blocking the expression of an action tendency without reducing the strength of the action tendency.

The main stumbling-block in Atkinson and Birch's theoretical analysis of the dynamics of action is the initial premise which they call the "principle of action". This states that the behaviour observed at any time is the expression of the dominant behavioural tendency, or the dominant combination of partially compatible tendencies. Thus the "observation of a change in activity implies a change in the dominance relations among the behavioural tendencies of the individual". (Atkinson and Birch, 1970.) The difficulty with this principle is that it assumes that all changes in activity are due to motivational competition (section 7.2.3; McFarland, 1969a). Atkinson and Birch (1970) would have to concede that manipulation of a second-in-priority tendency should affect the frequency of occurrence of the relevant behaviour according to their theory. It has been demonstrated, however, that this is not always the case (McFarland, 1969a, 1970b), and the possibility that a change in behaviour is due to disinhibition must therefore be allowed. The principle could possibly be salvaged by postulating a one-to-one relation between behavioural tendency and observed behaviour. This reduces the principle to a tautology, hardly a good start for the basic premise of a complex theoretical analysis.

A more empirical approach relating decision theory and motivation is that of Logan (1960, 1964, 1965a, b). Logan (1964) distinguishes between free and non-free behavioural situations. Conventional studies have shown that performance is generally an increasing function of drive (e.g. as a function of time of food deprivation) and of incentive (e.g. as a function of amount of food reward). In a free situation, where the subject controls his own drive, increased reward should produce greater incentive implying a higher frequency of response. On the other hand, increased reward should provide a larger intake and the resulting decrease in drive should imply a lower frequency of response. Since these two effects work in opposite directions the overall effect of increasing the amount of reward cannot be predicted without a quantitative formulation which takes account of the mutual interdependence of drive and performance in a free behaviour situation. In the free situation the subject must work on some specifiable "terms" for his ration of some commodity. Unless the terms are so hard that even continuous work cannot provide sufficient, the animal controls its own level of drive. In contrast to typical research on motivation where drive is the independent variable whose effects on performance are studied, in a free behaviour situation drive is also a dependent variable affected by various aspects of the terms.

Logan (1960) had rats living in a box in which they had to work for any water they received by pressing a bar. The terms were manipulated by varying the force required to depress the bar, the number of times the bar had to be pressed to obtain a reward and the size of the reward. He found that rats were sensitive to the details of the terms, in that daily water intake was reduced when any of the three aspects of the terms was increased. In subsequent experiments Logan (1964) was able to specify more precisely the relationship between terms and intake and some of his results are illustrated in Fig. 7.24. On the basis of such data, Logan (1964) proposes a

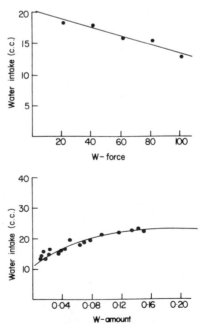

Fig. 7.24. Water intake as a function of the force aspect of the water terms (above), and of the amount aspect of the water terms (below), food being delivered automatically. (From Logan, 1964.)

dynamic model of free behaviour, which is formulated in Hull-Spence terms (Hull, 1952; Spence, 1956; Logan, 1960; see also section 7.1.1). The model assumes two thresholds; a higher value of response tendency (sEr) being required to initiate consumption of food (or water), which is then maintained until a lower value is reached that stops consumption. This type of phenomenon, generally called a "relaxation oscillation", is one that is commonly assumed to underlie behavioural aspects of homeostasis. Threshold phenomena involved in the initiation of behaviour patterns have been clearly identified in a number of cases. For example, there is evidence that systemic thirst factors elicit drinking only above a certain threshold (Wolf, 1958; Fitzsimons, 1963; McFarland and McFarland, 1968), and threshold devices having hysteresis have been incorporated into a number of models of the drinking control system (Reeve and Kulhanek, 1967; Oatley, 1967; Toates and Oatley, 1970). The first threshold applies to primary drinking but not to secondary drinking (see section 6.3.2). In the *ad libitum* situation much of the drinking is of the latter type (Kissileff, 1969; Fitzsimons and Le Magnen, 1969), and data on meal patterns suggests that a relaxation oscillation may be involved in the regulation of eating (Thomas and Mayer, 1968; Duncan *et al.,* 1970), but not of drinking (McFarland, 1969a). Thus although threshold phenomena may be relevant in some cases, they cannot be assumed to have universal importance. Moreover, the type of theory which relies on the existence of thresholds to account for the patterning of observed behaviour runs into the problem of what happens when more than one behavioural tendency is above threshold. This problem is recognized by Logan (1964)

". . . we will have to make explicit the assumption that performance is always a choice between the response in question and alternative behaviours. Indeed, the concept of threshold will probably give way to this response competition analysis; that is to say, we will view the threshold as the strengths of other possible responses."

This formulation is open to the objection, applicable to all theories which rely exclusively upon competition as the ultimate factor in the decision-making process, that it does not allow for the phenomenon of disinhibition. Nevertheless, Logan's work is important in that it focuses on the choice aspect of motivation, a neglected area in animal psychology. Logan's later work, especially, does much to bring the concepts of decision theory into the study of animal behaviour. Logan (1965a) gave rats a choice between a large food reward following a small delay, and a small reward following a large delay. He was able to establish equivalent combinations that were fit by "indifference" functions, and by equations describing the relative incentive value of different amounts and delays. In general, Logan's results were similar to those that would be expected by classical decision theory. Logan (1965b) gave rats a choice between a constant food reward and a varied amount, or varied delay, of reward; thus introducing the concept of "subjective probability" for the rat. The choice behaviour indicated that the net incentive value of varied reward is neither the average of the rewards nor the average of the incentives as independently estimated. The decision-making process in rats appears to be more complex than might have been thought.

A very different approach to decision making in animals is that of Dawkins (1969a). He notes that,

"Ethologists attempting to explain short-term switching between alternative behaviour patterns have often made use of a threshold type of model (Lorenz, 1950; Bastock and Manning, 1955; van Iersel and Bol, 1958; Andrew, 1961; von Holst and von Saint Paul, 1963; Tugendhat Gardener, 1964), in which the different items of behaviour are thought to be activated by the same 'drive', 'excitation' or equivalent, but at different threshold levels. These models were proposed to account for decision-making between alternative 'motor patterns'. The present model uses the same principle to explain decisions between alternative external stimuli to which a given response may be directed.

A threshold model was plausible to Bastock and Manning for *Drosophila* courtship partly because of the way in which three behaviour patterns were 'superimposed' on each other. Thus orientation, the 'low threshold' activity, could occur by itself but vibration only occurred if orientation also did; and licking, with the 'highest threshold', only occurred superimposed on vibration and orientation. If we are to make a similar model for choice, with the different stimuli having different thresholds, it is clearly impossible to consider an exactly parallel superimposition; a chick cannot peck simultaneously at two different targets. One could however imagine a probabilistic equivalent; instead of pecking simultaneously at the two targets the chick could go into a state of being equally likely to peck either of them. This is the basis of the choice threshold model now described."

Dawkins' model is fundamentally probabilistic in nature (see section 5.2.1), in that it incorporates a randomly fluctuating variable as part of the essential mechanism (Fig. 7.25). It is difficult to see in what way the model can "explain decisions". Essentially Dawkins passes a random waveform through a non-linear filter and it seems likely that almost any pattern of behaviour could be reproduced in the statistical sense, by judicious selection of the probability density function of the random waveform and of the characteristics of the filter. Nevertheless, there is no doubt that predictions of a statistical nature (i.e. concerning proportions of different types of choice), do

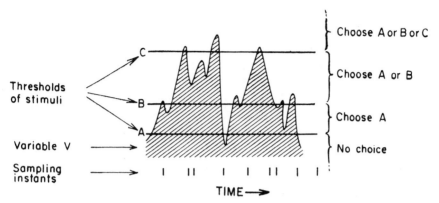

Fig. 7.25. Dawkins' choice threshold model. Three stimuli, A, B, C, preferred in that order, are considered. Variable V, drawn arbitrarily, interacts with thresholds corresponding to the stimuli. At "sampling instants" whose frequency is independent of V, choices are made according to the rule that probability of being chosen is equally divided between all stimuli whose thresholds are at that instant exceeded by V. (From Dawkins, 1969a.)

show good agreement with data obtained from a variety of animal experiments. Moreover, refinements of the basic model (Dawkins, 1969b; Dawkins and Impekoven, 1969) have extended its predictive power even further. But the question of the relationship between the type of explanation offered by a probabilistic, as opposed to a noisy deterministic, model remains.

7.3.2 Optimization and Motivation Theory

The importance of optima has long been recognized in biology (Rosen, 1967) and physiology (Milsum, 1968), but there has been little attempt to apply optimal control theory to behavioural problems. Of particular importance in motivation are the optimal conditions of the internal environment, with respect to blood chemistry, pH, temperature, etc. The role of behaviour in the maintenance of homeostasis has already been discussed at some length (sections 3.3.0 and 6.3.2) but one specific problem remains.

Suppose that an animal is kept without food and water, so that there are deviations from the optimal conditions of the internal environment with respect to a number of systemic variables. For the sake of illustration, these will be reduced to two, food deficit and water deficit. In reality more than two variables would be affected, though some of these might be additive in their effects on feeding or drinking behaviour (section 6.2.3). The problem is what is the optimal strategy for bringing these variables back to their normal values, given that the two relevant activities, feeding and drinking, cannot be carried out simultaneously?

By regarding the food deficit D_f and water deficit D_w as primary state variables in a motivational system (section 7.2.1), the condition of the animal after deprivation can be represented by a point in a hunger-thirst state plane (Fig. 7.26a) where zero is regarded as the

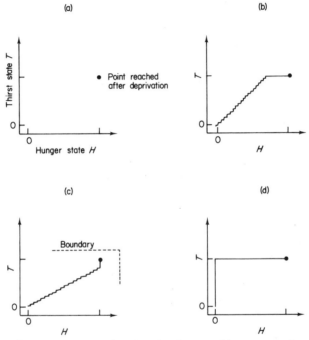

Fig. 7.26. Possible trajectories in the hunger-thirst state plane. (a) The condition at the end of deprivation of both food and water. (b) Recovery trajectory based upon minimization of the larger deficit (competition principle). (c) Recovery trajectory based upon avoidance of boundaries. (d) Recovery trajectory characterizing extreme persistence.

normal and optimal value for the two state variables. The problem now is to find the optimal trajectory for returning the initial (post-deprivation) state to zero. The answer to this question will largely depend upon the nature of the performance criteria. For example, if the overriding criterion is to minimize the larger deficit at any time, then the trajectory illustrated in Fig. 7.26b would probably be optimal. On the other hand, if it is more important to avoid certain critical values of the deficits, beyond which the animal will die, then the optimal trajectory might be that illustrated in Fig. 7.26c. Clearly, as we do not know what the performance criteria are we cannot specify the optimal trajectories *a priori*. This type of problem is therefore the inverse of the type of problem met in optimal control theory. We can assume that an optimal strategy is used, determine the trajectories empirically and then attempt to find the relevant performance criteria. Alternatively, we can postulate certain performance criteria on the basis of some general theory, and compare the empirical trajectories with those dictated by the theory. But we cannot take the performance criteria as given.

A view that is often implicit, and sometimes explicit (e.g. Atkinson and Birch, 1970), in motivation theory is that the activity observed at any particular time is the expression of the strongest of the current behavioural tendencies. Given equal weighting for the contribution of relevant external stimuli, this theory will produce a trajectory similar to that illustrated in Fig. 7.26b. That is, the animal responds first to its strongest motivational state variable, and behaves accordingly until the value of this variable is reduced below that of the next strongest, at which point the animal will switch activities. The exact point of the changeover will depend upon the degree of hysteresis in the system. Quite apart from empirical considerations this view has the disadvantage discussed in section 7.2.3. It tends to induce dithering, as illustrated by the frequent changes in the direction of the trajectory illustrated in Fig. 7.26b. A way out of this theoretical difficulty is to suppose that, having chosen a particular course of action, the animal locks on to it until the task is complete. The necessity for this type of mechanism has been recognized by a number of workers, and the phenomenon is often called "persistence" (Bolles, 1967; Atkinson and Birch, 1970). Extreme persistence would produce the type of trajectory illustrated in Fig. 7.26d. Few experiments have been made on this aspect of choice behaviour, but some preliminary experiments carried out in the author's laboratory suggest that animals exhibit neither complete dithering nor complete persistence. Some of this work will now be described.

The coordinates of a hunger-thirst state plane can be scaled by reference to arbitrarily chosen units of measurement, provided that the scaling remains consistent from one manipulation to the next. A convenient measure for the state variables involved in hunger and thirst is the percentage deviation from the normal (steady-state) value (McFarland, 1965c; Toates and Oatley, 1970). McFarland (1972) used a number of lines of evidence to estimate that the water deprived Barbary dove suffers a food deficit equivalent to one-eighth of that experienced during the same period of food deprivation. This enables the motivational state of the animal to be plotted as a trajectory in a hunger-thirst state plane, as illustrated in Fig. 7.27. During food deprivation Barbary doves remain in water balance (McFarland, 1965a, 1972) so that this trajectory moves directly along the food axis (Fig. 7.27). During simultaneous food

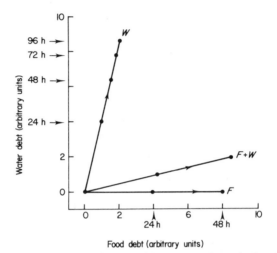

Fig. 7.27. Trajectories in the hunger-thirst state plane during deprivation of water (*W*), food (*F*), and food plus water (*F* + *W*). (From McFarland, 1972.)

and water deprivation, hunger rises at the same rate as during food deprivation, but the increase in thirst is attenuated by the reduced food intake, and the water debt after 48 h food-water deprivation is equivalent to that after 12 h water deprivation (McFarland, 1972). The trajectory for simultaneous food-water deprivation is correspondingly less steep (Fig. 7.27). McFarland (1972) conducted experiments in which doves were first deprived of water and then of food, at a specified time later. After a further specified period, choice tests between food and water were conducted to determine the relative preference.

The subjects used in these experiments were tested in a two-key Skinner box in which one key was illuminated with a red light and the other with a green light. For each peck at the green key 0.1 cc water was delivered and for each peck at the red key the subjects obtained three grains of wheat. After each peck the key lights were turned off, and the keys made inoperative for 30 s while the house lights remained on. The birds soon learned not to peck the keys when they were unilluminated. Each subject effectively received a food-water choice every 30 s, and each test session lasted for 30 min. Control tests showed that the choice measures employed were sensitive to a wide range of deprivation levels. Similar findings were obtained from choice tests given in a Grice box as a check on the validity of the choice-test procedure (McFarland, 1972). It was found that the relative food-water preferences were the same for deprivation schedules which had the same resultant vector on the hunger-thirst plane (vector 0Z in Fig. 7.28). This study shows that description of motivational states in terms of vectors is a realistic proposition and that the choice measures employed provide a reasonable indication of primary motivational state, at least with respect to hunger and thirst.

During food or water deprivation, the point in the state space, representing the primary motivational state, moves away from the

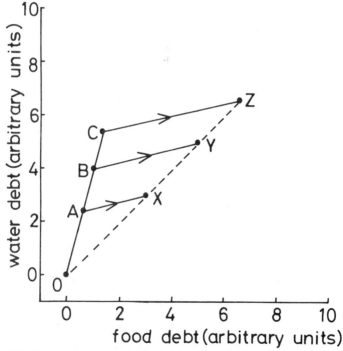

Fig. 7.28. Deprivation trajectories obtained by depriving doves firstly of water alone, and then, at a specified time later, of food also for a further period. Animals given choice tests at points X, Y and Z, showed equivalent food-water preferences. (From McFarland, 1972.)

origin. Although its rate of progression can be slowed down by the action of physiological conservation mechanisms, it cannot reverse direction without behaviour being involved. The change in primary motivational state brought about by feeding and drinking behaviour can be measured in terms of the quantities ingested and it has been found that Barbary doves (McFarland, 1972), like some other species (Adolph, 1950), are remarkably accurate in ingesting the quantity needed to restore a deficit to zero. However, the primary state variables [x] are not affected immediately, as there is some delay involved in absorption (see section 3.3.2) and evaluation of motivational state by integration of behavioural flow variables is only valid as a long-term solution. From the behavioural point of view, the action of short-term satiation mechanisms effectively overrides the delay and dictates behaviour except when systemic changes occur during the process of ingestion. For example, when thirsty doves are allowed to drink saline, the concentration of which is below the taste-rejection threshold, they drink the same quantity as distilled water, unless the ingestion rate is limited to the extent that some saline is absorbed before ingestion has finished. In such cases more saline than water is consumed (see section 5.1.4).

In general terms (Fig. 7.19), behaviour is determined by the command vector $[e_c]$. But, in the case of recovery from simultaneous food and water deprivation, two such commands are relevant and they can be carried out only one at a time. It is therefore of interest to determine the trajectories characteristic of this type of behavioural situation by employing the type of choice test described above.

Figure 7.29 shows averaged results obtained from six doves given half-hour choice tests during recovery from food and/or water deprivation. The number of food and water choices is plotted cumulatively for successive 5-min intervals. The following points arise from these results: (1) the curves are linear, suggesting that the choices are made on the basis of a persisting relationship between the hunger and thirst command variables. These might be expected to change little throughout the duration of the test, in view of the small quantities of food and water ingested. (2) Although food deprived doves remain in water balance some water choices are made by doves deprived of food only (curve $F24$). This is not altogether surprising as it is known that food intake to some extent dictates water intake (see section 5.2.3). (3) The curve for doves deprived of water for 48 h and food for 24 h (curves $W48$, $F24$) takes a route which suggests that the animal takes a "compromise" course between the

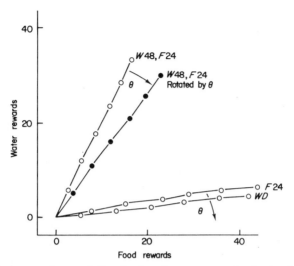

Fig. 7.29. Averaged cumulative food and water rewards obtained by six doves in choice tests given after specified periods of deprivation. (From McFarland, 1972.)

two commands. (4) The curve for doves deprived of water for 48 h and food for 24 h, but allowed to drink freely immediately before the test is rotated from its origin towards the food axis, and lies close to the $F24$ curve. This supports the view that the trajectories are based on a relationship between command vectors, since the initial water intake has an immediate effect which presumably operates via short-term satiation mechanisms. The view that the thirst component of the $F24$ curve is due to food-dictated water intake is also supported by this result because the degree of water choice is equivalent to that after food deprivation alone, even though thirst had just been satiated. (5) It is likely that every test is affected by food-dictated water intake, and each should therefore be rotated through angle θ, to correct for this artefactual effect of the testing procedure.

If the choice curves are based upon the relative levels of hunger and thirst commands, a shift in the command relationship during ingestion should cause the curve to deviate from its initial path. This prediction was investigated in the following experiment. A group of six doves were given two pairs of 30-min choice tests, of the type described above. Each test was conducted after 48 h water plus 24 h food deprivation. In the first pair of tests the birds were allowed to drink freely immediately before the test. In the second pair a 30 min delay was inserted between the drink and the start of the choice test. Each test pair consisted of a comparison of choices made after a prior drink of either 0.5% saline or distilled water. The treatments were carried out on a latin square basis so that each bird acted as its own control. The concentration of saline used is below the rejection threshold and the mean quantities of saline and distilled water drunk were always the same.

The results, illustrated in Fig. 7.30, show that when the birds were

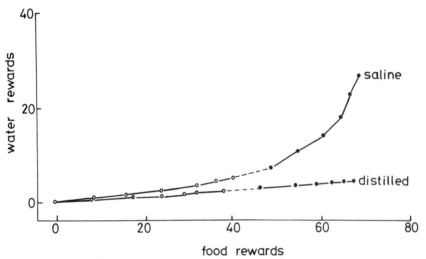

Fig. 7.30. Cumulative rewards obtained in choice tests following prior administration of saline or distilled water. (From McFarland, 1972.)

given distilled water to drink immediately (open circles) before the test the choice curve is much the same as that obtained from previous similar experiments (curve *WD, F*24, Fig. 7.29). When given distilled water after a 30-min delay (black circles), the curve is essentially the same, and can be plotted as a continuation of the non-delayed case, since the duration of the delay is the same as that of the non-delayed test. When given saline, however, the non-delayed curve (open circles) deviates from that of distilled water and the deviation continues in the same direction when the birds are tested 30 min after drinking saline (black circles). These results suggest that ingestion of saline causes a shift in food-water choice, due to absorption of saline into the bloodstream (see also section 5.1.4). The finding that the delayed curve is a continuation of the non-delayed curve suggests that the shift in choice is due only to the consequences of absorption and not to behavioural factors.

An important conclusion from these experiments is that food-water choices are made on the basis of the relative strengths of the hunger and thirst commands, which can be identified with the generalized effort vectors [e_c] (Fig. 7.19). The evidence for this conclusion may be summarized as follows: the linearity of the choice curves suggests that the choice is made on the basis of a persistent relationship between the actuating variables. The finding that the direction of choice is immediately affected by allowing the birds to drink before the test shows that the actuating variables are subject to short-term satiation factors. The manner in which food-water choice is influenced by saline ingestion suggests that the actuating variables are also influenced by systemic factors. The results of the choice experiments can be described in a hunger-thirst state plane by translating the primary state variables (deficits) into grams of food and water necessary to bring the motivational state to zero (i.e. the steady-state or satiation level). The state at the start of each experiment is then marked at the appropriate point in the state plane (Fig. 7.31) and the amounts (grams) of food and water consumed in

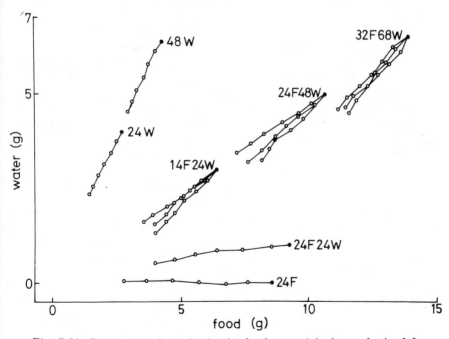

Fig. 7.31. Recovery trajectories in the food-water (q) plane, obtained from choice tests following specified periods of deprivation. (From McFarland, 1972.)

each successive 5 min period of the test are subtracted from each axis, after a correction (see above) has been made for food-induced water intake. In this manner six points on a trajectory are obtained (Fig. 7.31). When plotted in this way the results of the choice experiments show that, when the ingestion rate is severely limited, the trajectory in the hunger-thirst state plane heads straight for the origin. It follows a "compromise" path from the outset, and does not initially minimize the stronger motivational state as illustrated in Fig. 7.26b. This finding suggests that the system is designed to minimize a "joint" function of the relevant motivational state variables.

Although there is some effect of food intake upon water intake, the main level of motivational interaction involved in these experiments is that of the behavioural final common path (section 7.2.3). In terms of Fig. 7.19, it is the inductive part of the system that is primarily involved. However, the choice-test procedure effectively precludes normal operation of this part of the system. The behaviour does not have its normal consequences and little behavioural momentum can develop. The results of further experiments show that the hunger-thirst trajectories are considerably affected by the rate of ingestion permitted by the experimenter.

When doves are allowed to work freely for food and water in a Skinner box they alternate between feeding and drinking, but not as rapidly as the food-water choice experiments might suggest. The procedure is essentially the same as that employed in the choice tests except that the time (time out) after each reward, when the keys are unilluminated and inoperative, is greatly reduced. Thus an animal may be able to obtain a reward every second, rather than every 30 s, as in the choice tests. A typical result is shown in Fig. 7.32, and an

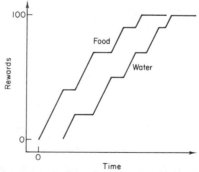

Fig. 7.32. Typical (idealized) cumulative record from an animal working for both food and water in a Skinner box.

important point is illustrated by this example. The animal tends to "lock-on" to food or water, taking fairly large amounts of each before changing.

A measure of the degree to which the animals lock-on to food or water is illustrated in Fig. 7.33. The "lock-on index" is essentially the sum of the shaded areas. If an animal took all its food before

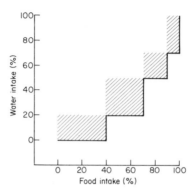

Fig. 7.33. Percentage of total water intake during a single session, plotted against that of food intake. A lock-on index is obtained by summing the shaded areas.

taking any water, the lock-on index would be maximal at 100, and if it alternated very frequently between food and water, the index would be low. Experiments (McFarland and Lloyd, 1971b) show that the lock-on index is a function of reward rate (Fig. 7.34). The reward rate is manipulated by altering the time-out after each reward. Thus with a time-out of 1 s the doves obtain 15 rewards/min on average. Control experiments show that it is rate of reward in terms of quantity of substance ingested that is important. When the reward size is varied in conjunction with reward rate, so that the quantity ingested per unit time is constant, the lock-on index also remains constant. Similarly, when the number of pecks required to maintain a given reward rate is varied the lock-on index remains the same. In other words, the degree of locking on is not affected by the rate of rewarding events, or by the amount of work required. It is altered only by the rate of obtaining the reward substance and the faster this is obtained the more the animal locks on to the appropriate behaviour.

These findings suggest that positive feedback from certain consequences of the behaviour is an important factor determining the degree of locking on: the less the feedback the more the animal

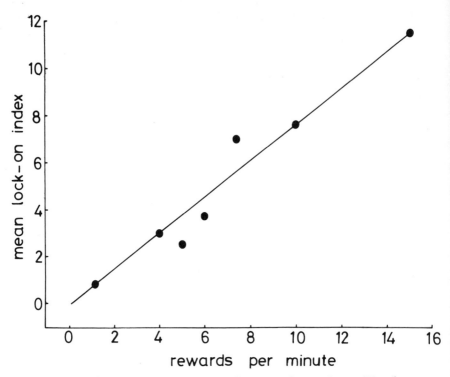

Fig. 7.34. Mean lock-on index as a function of reward rate. ($N = 6$).

switches to alternative behaviour. In terms of control theory, positive feedback would be expected to increase the momentum of the behaviour (section 7.2.1), and it has been suggested (section 3.3.3) that in doves oral factors are primarily responsible for reinforcing ongoing behaviour and maintaining the momentum. However, it appears that reward rate is not the only factor determining the degree of locking on. If a transparent partition is introduced between the food and water keys (Fig. 7.35), and the reward rate is held constant (3 s time-out), the lock-on index increases as a function of the partition length (Fig. 7.36). The food satiation curves are not affected by the partition length, as the bird has plenty of time to walk around the partition during the time-out period. The amount of water ingested during the session is also not affected. Thus the only apparent effect of the partition is to increase the lock-on index. It is as if the animal weighed up the "cost" of changing its behaviour before deciding to change. Further experiments are necessary to discover what factors are important in this phenomenon but it seems

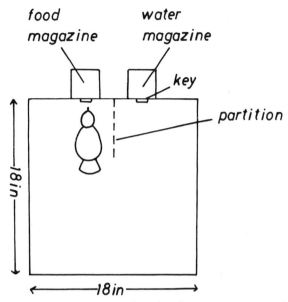

Fig. 7.35. Plan view of testing chamber showing transparent partition between the food and water keys.

clear that the state of the environment has an effect upon the behavioural momentum.

7.3.3 Causal and Functional Aspects of Optimization

In general, the results of the experiments outlined in the previous section suggest that an overall aim, of minimizing a joint function of the values of motivational state variables at any time, is offset by the rate of progress of minimization of any one variable. In other words, when reward rate is high behaviour relevant to other motivational systems tends to be postponed. In addition it appears probable that the "cost" of changing from one activity to another is also taken into account in the optimal strategy. However, the situation is further complicated by the fact that behaviour occurring as an adjunct to ongoing feeding or drinking behaviour occurs by disinhibition (McFarland, 1970b). That is, the magnitude of the second-in-priority motivational state variable need have no effect upon the patterning of the top-priority behaviour, even though the second-in-priority behaviour may frequently appear to interrupt the top-priority behaviour. In terms of the state plane, the points at which the

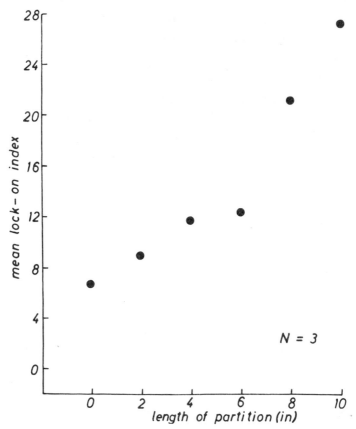

Fig. 7.36. Mean lock-on index as a function of partition length ($N = 3$).

trajectory is diverted from the dominant path (i.e. the path first chosen by the animal) is not affected by the magnitude of the motivational state in any other dimension. An example of this phenomenon is illustrated in Fig. 7.37.

As a working hypothesis, the strategy followed by the animal may be summarized as follows: (1) weight each primary motivational state vector in relation to the prevailing external stimulus situation. (2) Choose the resultant with the largest magnitude and initiate the appropriate behaviour (up to this point the procedure could be much the same as that postulated by classical decision theory). (3) Monitor the consequences of the ongoing behaviour and if satisfactory (by some set of criteria) continue. If not satisfactory, consider switching to an alternative behaviour (generally that relevant to the

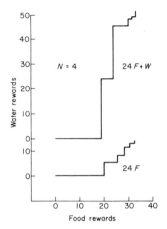

Fig. 7.37. Cumulative food and water rewards averaged from animals tested after food deprivation (lower) and food plus water deprivation (upper). (Data from McFarland, 1970b.)

second-in-priority motivational state). (4) Decide to switch to an alternative activity on the basis of a combined function of the satisfactoriness of the ongoing behaviour, and the "cost" of switching.

This hypothesis has much in common with McFarland's (1966a) "frustration diverts attention" theory of the disinhibition of displacement activities (section 7.2.3), and with Baerends' (1970) explanation of the occurrence of interruptive behaviour during incubation in the Herring gull (section 7.1.2). In effect, the hypothesis states that animals do not always respond to the "strongest" current motivation or behavioural tendency, but may "time-share" between a number of motivational tendencies in accordance with certain rules. The assumption underlying the hypothesis is that the rules conform to some optimality principle. The problem of finding the rules that govern this time-sharing optimization is the inverse of that normally found in optimization theory, and in this respect it is important to distinguish between causal and functional aspects of optimization.

When an engineer designs an optimizing system he does so with a particular environment in mind. In choosing performance criteria (in relation to that environment) he is attempting to achieve certain design aims and this activity can be seen as an attempt to optimize a particular function. In other words, design can itself be seen as an optimizing activity. The engineer designs the system to perform a certain function and in doing so he decides what performance criteria are relevant under the circumstances. The selection of performance criteria, and other aspects of the design activity, may be regarded as "functional" aspects of the optimizing system. The performance criteria involved in this type of activity, the design goals, may be very different to those embodied in the system itself, which may be called the "causal" aspects of optimization. Engineers normally take the functional aspects of optimization for granted and are not concerned to analyse them.

In the biological world, design is exercised by the process of natural selection, itself an optimizing system (Rosen, 1967). A biological system may be designed by natural selection to fulfil a certain function in relation to a particular environment. For example, it may be functionally important for a Herring gull to produce eggs of a certain (optimal) size. The manner in which the system fulfils that function may involve very different optimization principles from those involved in the design. Thus the Herring gull does not prefer eggs of the optimal size, but prefers a larger, "supernormal" egg, when given a choice. The normal environment, however, involves constraints such that these large eggs are not available. The existence of such constraints ensures that the functional performance criteria are attained by a system which involves very different (causal) performance criteria and optimizing principles. It may indeed be more "efficient" to design a system involving a "bigger-the-better" principle (given the environmental constraints), than to design a system with a "preference" that matches the design criterion.

In the case of behavioural time-sharing, two types of question must be distinguished. On the one hand, one can ask what function time-sharing is designed to fulfil, what performance criteria are involved? On the other hand, there is the question of what criteria are used by the animal in the control of this type of behaviour, and what principles of optimization are involved in the causal sense. On the latter question little more can be said without further experimentation. At present the evidence for time-sharing obtained in the author's laboratory is limited to two species (doves and rats) and two types of motivational interaction. Both the generality of the phenomenon, and the rules involved in the behaviour, require further empirical investigation.

The functional problem seems to be more of a matter for theoretical investigation and it may be interesting to consider some of the possible lines of approach. The existence of displacement vectors [x], corresponding to physiological imbalance in a homeostatic motivational system (eqn. 7.3), represents a potential danger to the animal in that physiological consequences, of the value of [x] exceeding a certain boundary value, may be lethal. The probability P_x of such a displacement exceeding the boundary value increases as [x] increases, since the opportunity for the animal to reduce [x] by appropriate behaviour is lessened. The probability P_b of any one of a number of displacements exceeding a boundary value is given by the equation

$$P_b = P_x + P_a - P_x P_a \qquad (7.15)$$

where P_x is the probability of a particular displacement exceeding a boundary and P_a the probability of another displacement doing so. Suppose that an animal has the opportunity to reduce P_x by a given amount in a given time *or* to reduce both P_x and P_a by half that amount in the same time. The latter course will always result in a lower P_b Thus

$$(P_x - \tfrac{1}{2}P_r) + (P_a - \tfrac{1}{2}P_r) - (P_x - \tfrac{1}{2}P_r)(P_a - \tfrac{1}{2}P_r) < (P_x - P_r) +$$
$$P_a - (P_x - P_r)P_a \qquad (7.16)$$

where P_r is the probability reduction that is possible in the time available. In general, when an animal time-shares between activities that reduce the probability of a boundary being exceeded an advantage is gained in terms of the overall probability of a lethal boundary being exceeded.

Another possible approach to functional aspects of optimization is to consider the motivational energy changes involved. In terms of the energy concepts outlined in section 7.2.2, a progressive increase in the values of motivational state variables [x], such as would occur during food or water deprivation, can be seen as an accumulation of potential energy (eqn. 7.8). The energy level attained will depend not only upon the values of the displacement variables [x], but also upon the values of the generalized capacitance parameters associated with each displacement (section 7.2.1). Each element of the capacitance matrix can be regarded as a measure of the displacement, calibrated in terms of its "meaningfulness" to the animal. In other words, the output of a capacitance, an effort variable, is an indication of the capacity of the system to tolerate the displacement. The energy level associated with each element (eqn. 7.6) can thus be regarded as synonymous with the danger of a lethal physiological imbalance. For example, for a small capacitance a small displacement might be lethal, while for a large one, a larger displacement would be tolerable.

In any motivational system, there will be more than one type of displacement and more than one capacitance element, relevant to the primary state of the system. It is for this reason that it is necessary to think of motivational displacement and capacitance as vector [x] and matrix [1/C] respectively. Moreover, because the displacements may not all be independent, the total potential energy associated with the capacitance matrix should be calculated in terms of generalized coordinates (section 7.2.2). A motivational system with two degrees of freedom, for example, might involve two generalized effort variables, each a function of only one generalized coordinate. Two capacitances would therefore be involved, and in terms of eqn. 7.9

$$E_p([q]) = \int_0^{q_1} \frac{1}{C_1} q_1 dq_1 + \int_0^{q_2} \frac{1}{C_2} q_2 dq_2 = \frac{1}{2C_1} q_1^2 + \frac{1}{2C_2} q_2^2 \qquad (7.17)$$

where C_1 and C_2 represent the two capacitances, and q_1 and q_2 the corresponding generalized coordinates.

On the basis of the above considerations, we might expect an animal to behave in a manner that minimizes potential energy as fast as possible. In other words: to minimize the time-integral of the potential energy, from the time that the opportunity is available. Generally, the opportunity for behaviour related to a number of motivational systems, such as hunger, thirst, sex, comfort, etc. will exist simultaneously. Thus it is the time-integral of the total potential energy that we should expect to be minimized, viz.

$$\text{minimize} \int_0^t (E_1 + E_2 + E_3 \ldots + E_n)\, dt \qquad (7.18)$$

where E_1 to E_n are the potential energy levels associated with those motivational states that the animal has the opportunity to reduce. On this basis it is a simple matter to predict the optimal course of action. The time-integral of the total potential energy will always be minimal if the animal follows a course of action that is relevant to the motivational system with the highest potential energy level. This functional principle will be followed by an animal whose actions are dictated by motivational competition (section 7.2.3), provided that decisions are made on the basis of motivational displacements, which have been "calibrated" in terms of the animal's capacity to tolerate each type of displacement. Changes in behaviour will occur as soon as the level of the current calibrated displacement (or equivalent command) falls below that of another. Thus an animal with large displacements in two dimensions will follow the course illustrated in

Fig. 7.26b. Such a course minimizes the total potential energy and may be said to minimize "liability" with respect to physiological consequences.

In considering an animal in its natural environment it is easy to see that minimization of liability alone is unlikely to be the functionally optimal strategy. Changes in behaviour determined solely on the basis of motivational competition do not allow for two important functional considerations. The first is the loss of opportunity that may be involved in such a change. For example, an animal may continue feeding up to the point where its hunger (command) is reduced to the level of its second-in-priority command, which may be sexual, for instance. Competition theory dictates that there will be a change from feeding to sexual behaviour at this point. Under natural conditions such a change might deprive the animal of the opportunity to exploit that particular food source. If, on the other hand, the animal were able to "finish the job", despite competing motivations, the opportunity might not be wasted.

The second functional consideration concerns the cost involved in changing from one activity to another. The feeding animal may have to travel a long distance to satisfy its sexual motivation, but as long as there are cues that indicate the possibility of such action competition theory implies that the change in behaviour will occur regardless of cost.

Functional considerations of opportunity and cost relate directly to ongoing behaviour, and its consequences. It may therefore be possible to formulate such considerations in terms of kinetic energy (section 7.2.2). Generally, kinetic energy will be high when "the going is good" and the ongoing behaviour develops a momentum capable of overriding competition from other motivational systems. The necessity for a mechanism of this type is discussed in section 7.2.3. On a similar basis we might expect that there would be a loss of kinetic energy when the cost of a change in behaviour is high.

In considering the process of decision making, maximization of kinetic energy would seem to be a good design principle. A high kinetic energy level will generally be correlated with a high rate of dissipation of potential energy and with a high behavioural momentum. A system designed to optimize a joint function of kinetic and potential energy might meet many of the (functional) performance criteria involved in the design, by natural selection, of the behavioural final common path. This possibility has much in common with several optimization methods which employ energy concepts. For example, the behaviour of a physical system can often be described in terms of Lagrangian dynamics (Wells, 1967), a type of formulation that forms the basis of a number of optimality principles in physics (e.g. Hamilton's Principle), and in systems analysis in general (e.g. the Maximum Principle of Pontryagin). In theoretical biology this type of approach is of increasing importance (Rosen, 1967), and in behaviour it seems promising. However, an essential first step in this direction must be a proper distinction between causal and functional aspects of optimization. In the above considerations, for example, the energy concept is used in an entirely functional sense. It is not envisaged that the animal measures motivational energy levels, or acts on the basis of any such information. Indeed, it is not envisaged that motivational energy plays any causal role (section 7.2.2). The value of energy formulations is primarily that they generate clear predictions, that can be tested experimentally and thus aid the complex task of unravelling the rules by which animals make decisions.

References and Author Index

The numbers in the square brackets refer to the page or pages in the text where mention of a given work or person is made.

Adolph, E. F. (1950). Thirst and its inhibition in the stomach. *Am. J. Physiol.* **161**, 374-386. [57, 177, 242]

Adolph, E. F. and Northrop, J. P. (1952). Absorption of water and chloride (Rat). *Am. J. Physiol.* **168**, 311-319. [62]

Amsel, A. (1958). The role of frustrative non-reward in non-continuous reward situations. *Psychol. Bull.* **55**, 102-119. [207]

Anand, B. K., China, G. S. and Singh, B. (1962). Effect of glucose on the activity of hypothalamic "feeding centres". *Science* **138**, 597-8. [56]

Andersson, B. (1953). The effect of injections of hypertonic NaCl solutions into different parts of the hypothalamus of goats. *Acta physiol. scand.* **28**, 188-201. [56]

Andrew, R. J. (1961). The motivational organisation controlling the mobbing calls of the blackbird. 2. Quantitative analysis of changes in the motivation of calling. *Behaviour* **17**, 288-321. [236]

Andrew, R. J. (1969). The effects of testosterone on avian vocalisations. In Hinde, R. A. (1969). [183]

Armstrong, E. A. (1950). The nature and function of displacement activities. *Sym. Soc. exp. Biol.* **4**, 361-384. [187]

Ashby, W. R. (1956). "An Introduction to Cybernetics." Chapman and Hall, London. [141]

Ashby, W. R. (1960). "Design for a Brain." 2nd Ed. Chapman and Hall, London. [22]

Aschoff, J. (1963). Diurnal rhythms. *A. Rev. Physiol.* **25**, 581-600. [146]

Atkinson, J. W. (1964). "An Introduction to Motivation". Van Nostrand, Princeton. [232]

Atkinson, J. W. and Cartwright, D. (1964). Some neglected variables in contemporary conceptions of decision and performance. *Psychol. Rep.* **14**, 575-590. [232]

Atkinson, J. W. and Birch, D. (1970). "The dynamics of action". John Wiley and Sons, New York and London. [232, 233, 239]

Audley, R. J. (1960). A stochastic model for individual choice behavior. *Pychol. Rev.* **67**, 1-15. [118]

Baerends, G. P. (1962). La reconnaissance de l'oeuf par le Goéland argenté. *Bull. Soc. Sci. de Bretagne* **37**, 193-208. [190]

Baerends, G. P. (1970). A model of the functional organisation of incubation behaviour. *In* "The Herring Gull and its Egg" (Baerends, G. P. and Drent, R. H., eds). *Behaviour,* **Suppl. XVIII**, 265-310. [203, 204, 251]

Baerends, G. P., Brouwer, R. and Waterbolk, H. Tj. (1955). Ethological studies on *Lebistes reticulatus* (Peters): I. An analysis of the male courtship pattern. *Behaviour* **8**, 249-334. [124, 188]

Bartholomew, G. A. and Cade, T. J. (1963). The water economy of land birds. *Auk.* **80**, 504-539.   [56]

Bastock, M. and Manning, A. (1955). The courtship of *Drosophila melanogasta. Behaviour* **8**, 85-111.   [236]

Bastock, M., Morris, D. and Moynihan, M. (1953). Some comments on conflict and thwarting in animals. *Behaviour* **6**, 66-84.   [201, 203]

Bayliss, L. E. (1966). "Living Control Systems." English Universities Press, London.   [13, 99]

Beer, C. G. (1961). Incubation and nestbuilding behaviour of black-headed gulls: I. Incubation behaviour in the incubation period. *Behaviour* **18**, 62-106.   [203]

Beer, C. G. (1965). Clutch size and incubation behaviour in black-billed gulls *(Larus bulleri). Auk.* **82**, 1-18.

Bellman, R. E. (1957). "Dynamic Programming." Princeton University Press.   [231]

Bendat, J. S. and Piersol, A. G. (1966). "Measurement and Analysis of Random Data." John Wiley and Sons, New York and London.   [119, 121, 122, 124, 126, 128, 130, 154, 164, 165]

Benzinger, T. H. (1969). Heat regulation: Homeostasis of central temperature in man. *Physiol Rev.* **49**, 671-759.   [102]

Bierens de Haan, J. A. (1947). Animal psychology and the science of animal behaviour. *Behaviour* **1**, 71-80.   [187]

Birmingham, H. P. and Taylor, F. V. (1954). A design philosophy for man-machine control systems. Proc. I.R.E. **62**, 1748-1758.   [109, 110]

Black, S. L. and Vanderwolf, C. H. (1969). Thumping behaviour in the rabbit. *Physiol. Behav.* **4**, 445-449.   [19]

Blass, E. M. and Fitzsimons, J. T. (1970). Additivity of effect and interaction of a cellular and an extracellular stimulus of drinking. *J. comp. physiol. Psychol.* **70**, 200-205.   [163, 175]

Bloomfield, T. M. (1968). Behavioural contrast and the peak shift. *In* "Animal Discrimination Learning" (Gilbert, R. M. and Sutherland, N. S., eds), pp. 215-240. Academic Press, London and New York.   [213]

Blurton-Jones, N. G. (1968). Observations and experiments on causation of threat displays of the great tit *(Parris major). Anim. Behav. (Monogr.)* 1.2.   [125]

Bolles, R. C. (1958). The usefulness of the drive concept. *In* "Nebraska Symposium on Motivation" (Jones, M. R., ed.). Nebraska University Press, Lincoln.   [223]

Bolles, R. C. (1961). The interaction of hunger and thirst in the rat. *J. comp. physiol. Psychol.* **54**, 58-64.   [116]

Bolles, R. C. (1967). "Theory of Motivation". Harper and Row, New York.   [187, 207, 209, 212, 213, 223, 239]

Boring, E. G. (1946). "A History of Experimental Psychology". Appleton-Century-Crofts, New York.   [137]

Brobeck, J. R. (1960). Food and temperature. *Recent Prog. Horm. Res.* **16**, 439-446.   [56]

Brockway, B. F. (1969). Roles of budgerigar vocalisation in the integration of breeding behaviour. In Hinde (1969).   [183]

Brown, B. M. (1965). "The Mathematical Theory of Linear Systems". 2nd Ed. Science Paperbacks. Associated Book Publishers Ltd., London.   [44, 119, 151, 152, 163]

Brown-Grant, K. (1966). Regulation and control in the endocrine system. *In* "Regulation and Control in Living Systems" (Kalmus, H., ed.). John Wiley and Sons, New York and London. [212]

Budgell, P. (1970). The effect of changes in ambient temperature on water intake and evaporative water loss. *Psychon. Sci.* **20**, 275-276. [69, 164]

Budgell, P. (1971). Behavioural thermoregulation in the Barbary dove *(Streptopelia risoria). Anim. Behav.* (In press). [75, 76]

Budgell, P. and McFarland, D. J. (1971). Thermal consequences of water ingestion (In preparation). [225]

Cade, T. J. (1964). Water and salt balance in granivorous birds. *In* "Thirst", Proceedings of the 1st International Symposium on Thirst in the Regulation of Body Water (Wayner, M., ed.). Pergamon Press, Oxford. [56]

Campbell, B. A. and Misannin, J. R. (1969). Basic drives. *A. Rev. Psychol.* **20**, 57-84. [224]

Campbell, F. W. and Robson, J. G. (1958). Moving visual images produced by regular stationary patterns. *Nature, Lond.* **181**, 362. [136]

Campbell, F. W., Robson, J. G. and Westheimer, G. (1959). Fluctuations in accommodation under steady viewing conditions. *J. Physiol.* **145**, 579-594. [133, 134, 135, 136]

Campbell, F. W. and Whiteside, T. C. D. (1950). Induced pupillary oscillations. *Br. J. Opthalmol.* **34**, 180-189. [93]

Carr, W. J. (1962). The effect of adrenalectomy upon the NaCl taste threshold in rat. *J. comp. physiol. Psychol.* **45**, 377-380.

Carr, W. J. and Caul, W. F. (1962). The effect of castration in rat upon the discrimination of sex odours. *Anim. Behav.* **10**, 20-27. [187]

Chapanis, A. (1951). Theory and method for analysing errors in man-machine systems. *Ann. N.Y. Acad. Sci.* **51**, 1179-1203. [116]

Chow, Y. and Cassignol, E. (1962). "Linear Signal-flow Graphs and Applications". John Wiley and Sons, New York and London. [12]

Chung, S. H. (1965). Effects of effort on response rate. *J. Exp. Anal. Behav.* **8**, 1-7. [138, 139, 213]

Cicala, G. A. (1961). Running speed in rats as a function of drive level and presence or absence of completing response trials. *J. exp. Psychol.* **62**, 329-334. [226]

Clynes, M. (1961). Unidirectional rate sensitivity. A biocybernetic law of reflex and humoral systems as physiological channels of control and communication. *Ann. N.Y. Acad. Sci.* **93**, 946-969. [87, 89, 90, 91]

Colby, K. M. (1955). "Energy and Structure in Psychoanalysis". Ronald Press, New York. [222]

Cooke, J. E. (1965). Human decisions in the control of a slow-response system. D. Phil. Thesis, Oxford University. [108, 110]

Corbit, J. D. and Luschei, E. S. (1969). Invariance of the rat's rate of drinking. *J. comp. physiol. Psychol.* **69**, 119-125. [175]

Cotes, J. E. and Meade, F. (1960). Energy expenditure and mechanical energy demand in walking. *Ergonomics* **3**, 97-119. [83, 84]

Cotton, J. W. (1953). Running time as a function of amount of food deprivation. *J. exp. Psychol.* **46**, 188-198. [226]

Coulson, A. E. (1969). "An Introduction to Matrices". Longman, Green and Co., London. [151]

Cox, D. R. and Lewis, P. A. W. (1966). "The Statistical Analysis of Series of Events". Methuen, London.   [119, 124]

Crosbie, E. J., Hardy, J. D. and Fessenden, E. (1961). Electrical analog simulation of temperature regulation in man. *I.R.E. Trans. Bio-Med. Electron.*, BME-8 (4).   [138, 169]

Cross, B. A. and Green, J. D. (1959). Activity of single neurones in the hypothalamus: effect of osmotic and other stimuli. *J. Physiol.* **148**, 554-569.   [56]

Darwin, C. (1872). "The Expression of the Emotions in Man and the Animals". University of Chicago Press, Chicago, 1965.   [19]

Dawkins, R. (1969a). A threshold model of choice behaviour. *Anim. Behav.* **17**, 120-133.   [236]

Dawkins, R. (1969b). The attention threshold model. *Anim. Behav.* **17**, 134-141.   [237]

Dawkins, R. and Impekoven, M. (1969). The 'peck/no-peck decision maker' in the black-headed gull chick. *Anim. Behav.* **17**, 243-251.   [237]

Delius, J. D. (1968). Color preference shift in hungry and thirsty pigeons. *Psychon. Sci.* **13**, 273-274.   [187]

Delius, J. D. (1969). A stochastic analysis of the maintenance behaviour of skylarks. *Behaviour* **33**, 137-138.   [154, 155, 164]

Denier van der Gon, J. J. and Thuring, J. P. (1965). The guiding of human writing movements. *Kybernetik* **2**, 145-148.   [114]

Dernier van der Gon, J. J. and Wieneke, G. H. (1969). The concept of feedback in motorics against that of preprogramming. *In* "Biocybernetics of the Central Nervous System" (Proctor, L. D., ed.). Little, Brown and Co., Boston, Mass.   [114]

Dicker, S. E. and Haslam, J. (1966). Water diuresis in the domestic fowl. *J. Physiol.* **183**, 225-235.   [62]

Douce, J. L. (1963). "An Introduction to the Mathematics of Servo-mechanisms". English Universities Press.   [130]

Duncan, C. J. (1960). Preference tests and the sense of taste in the feral pigeon. (*Columbia livia* Var Guselin). *Anim. Behav.* **8**, 54-60.   [66, 112]

Duncan, I. J. H., Home, A. R., Hughes, B. O. and Woodgush, D. G. M. (1970). The pattern of food intake in female brown leghorn fowls as recorded in a Skinner box. *Anim. Behav.* **18**, 245-255.   [120, 125, 127, 235]

Edwards, W. (1954). The theory of decision making. *Psychol. Bull.* **51**, 380-417.   [229, 232]

Edwards, W. (1961). Behavioural decision theory. *A. Rev. Psychol.* **12**, 473-498.   [229]

Edwards, W. and Tversky, A. (1967). Decision Making (Foss, B. M., ed.). Penguin Modern Psychology Readings.   [229]

Elgerd, O. I. (1967). Control Systems Theory. McGraw-Hill, New York. [45, 151, 231]

Epstein, A. N. (1960). Water intake without the act of drinking. *Science* **131**, 497-498.   [68]

Epstein, A. N. and Teitelbaum, P. (1962). Regulation of food intake in the absence of taste, smell, and other oropharyngeal sensations. *J. comp. physiol. Psychol.* **55**, 753-759. [68]

Epstein, A. N., Fitzsimmons, J. T. and Rolls, B. J. (née Simons) (1970). Drinking induced by injection of angiotensin into the brain of the rat. *J. Physiol.* **210**, 457-474. [159, 225]

Estes, W. K. and Burke, C. J. (1953). A theory of stimulus variability in learning. *Psychol. Rev.* **60**, 276-286. [116]

Falk, J. L. (1969). Conditions producing psychogenic polydipsia in animals. *Ann. N.Y. Acad. Sci.* **151**, 569-593. [227]

Falls, J. B. (1969). Functions of territorial song in the white-throated sparrow. In Hinde (1969). [183]

Feather, N. T. (1963). Mowrer's revised two-factor theory and the motive-expectancy-value model. *Psychol. Rev.* **70**, 500-515. [232]

Fitzsimons, J. T. (1958). Normal drinking in rats. *J. Physiol.* **138**, 39. [127]

Fitzsimons, J. T. (1963). The effects of slow infusions of hypertonic solutions on drinking and drinking thresholds in rats. *J. Physiol.* **167**, 344-354. [175, 235]

Fitzsimons, J. T. (1968). La soif extracellulaire. *Annls Nutr. Aliment.* **22**, 131-144. [209]

Fitzsimons, J. T. (1969). Effect of nephrectomy on the additivity of certain stimuli of drinking in the rat. *J. comp. physiol. Psychol.* **68**, 308-314. [163, 175]

Fitzsimons, J. T. and Le Magnen, J. (1969). Eating as a regulatory control of drinking in the rat. *J. comp. physiol. Psychol.* **67**, 273-283. [127. 160, 227, 235]

Fitzsimons, J. T. and Oatley, K. (1968). Additivity of stimuli for drinking in rats. *J. comp. physiol. Psychol.* **66**, 450-455. [163, 175]

Freud, S. (1932). "New Introductory Lectures on Psychoanalysis". Hogarth Press, London. [220]

Freud, S. (1940). "An Outline of Psychoanalysis". Hogarth Press, London. [215, 220, 221, 222]

Gann, D., Schoeffler, J. D. and Ostrander, L. (1968). A finite state model for the control of adrenal corticosteroid secretion. In "Systems Theory and Biology" (Mesarovic, M. D., ed.). Springer-Verlag, Berlin. [212]

Garcia, J., Erwin, F. R. and Koelling, R. A. (1966). Learning with prolonged delay of reinforcement. *Psychon. Sci.* **5**, 121-122. [114]

Garcia, J., Erwin, F. R., Yorke, C. H. and Koelling, R. A. (1967). Conditioning with delayed vitamin injections. *Science* **155**, 716-718. [114]

Gibb, J. (1954). Feeding ecology of tits, with notes on treecreeper and goldcrest. *Ibis* **96**, 513-543. [226]

Gibbs, C. B. (1970). Servo-control systems in organisms and the transfer of skill. In "Skills" (Legge, D., ed.). Penguin Modern Psychology Readings. [114]

Gilbert, E. G. (1963). Controllability and observability in multivariable control systems. *J. Soc. ind. appl. Math.* (ser. A, Control) **1**, 128-151. [145]

Goldman, S. (1961). Further considerations of cybernetic aspects of homeostasis. *In* "Self Organising Systems". Pergamon Press, New York and Oxford. [107]

Gray, J. A. and Smith, P. T. (1969). An arousal-decision model for partial reinforcement and discrimination learning. *In* "Animal Discrimination Learning" (Gilbert, R. M. and Sutherland, N. S., eds). Academic Press, London and New York. [118]

Green, D. M. and Swets, J. A. (1966). "Signal Detection Theory and Psychophysics". John Wiley and Sons, New York and London. [118]

Gregory, R. L. (1966). "Eye and Brain. The Psychology of Seeing". Weidenfeld and Nicolson, London. [198]

Grodins, F. S. (1959). Integrative cardiovascular physiology: A mathematical synthesis of cardiac and blood vessel hemodynamics. *Q. Rev. Biol.* **34**, 93-116. [138, 169]

Grodins, F. S. (1963). "Control Theory and Biological Systems". Columbia University Press. [169]

Hammel, H. T. (1965). Neurones and temperature regulation. *In* "Physiological Controls and Regulations" (Yamamoto, W. S. and Brobeck, J. R., eds). W. B. Saunders Co., Philadelphia and London. [24, 56]

Hammond, P. H., Merton, P. A. and Sutton, G. G. (1956). Nervous gradation of muscular contraction. *Br. med. Bull.* **12**, 214-218. [25]

Hardy, J. D. (1965). The 'set-point' concept in physiological temperature regulation. *In* "Physiological Controls and Regulations" (Yamamoto, W. S. and Brobeck, J. R., eds). W. B. Saunders Co., Philadelphia and London. [23, 24]

Harter, M. R. and White, C. T. (1968). Periodicity within reaction time distributions and electromyograms. *Q. Jl. exp. Psychol.* **20**, 157-166. [118]

Hartley, M. G. (1962). An introduction to electronic analogue computers. Science Paperbacks. Methuen, London. [168]

Heiligenberg, W. (1965). The effect of external stimuli on the attack readiness of a cichlid fish. *Z. vergl. Physiol.* **49**, 459-464.

Heiligenberg, W. (1966). The stimulation of territorial singing in house crickets *(Acheta domesticus). Z. vergl. Physiol.* **53**, 114-129. [189, 191, 192]

Heiligenberg, W. (1969). The effect of stimulus chirps on a cricket's chirping *(Acheta domesticus). Z. vergl. Physiol.* **65**, 70-97. [191]

Hein, A. and Held, R. (1962). A neural model for labile sensori-motor coordinations. *In* "Biological Prototypes and Synthetic Systems", 1. Plenum Press, New York. [198, 199]

Held, R. (1961). Exposure-history as a factor in maintaining stability of perception and coordination. *J. nerv. ment. Dis.* **132**, 26-32. [198]

Held, R. and Freeman, S. J. (1963). Plasticity in human sensorimotor control. *Science* **142**, 455-462. [199]

Held, R. and Hein, A. (1958). Adaptation of disarranged hand-eye coordination contingent upon re-afferent stimulation. *Percept. Mot. Skills Res.* **8**, 87-90. [199]

Held, R. and Hein, A. (1963). Movement-produced stimulation in the development of visually guided behavior. *J. comp. physiol. Psychol.* **56**, 872-876. [199]

Helmholtz, H. von (1867). Handbuch der physiologischen Optik. 1st Ed. Voss, Leipzig. [17, 197]

Helmholtz, H. von (1924). "Treatise on Physiological Optics". Vol. 1, p. 191. Menasha, Optical Society of America. [136]

Helmholtz, H. von (1962). "Physiological Optics". (Translated by J. P. C. Southall). Dover, New York. [17]

Hess, E. H. (1956). Space perception in the chick. *Scient. Am.* **195**, 71-80. [96, 97]

Hilgard, E. R. (1956). "Theories of Learning". 2nd Ed. Appleton-Century-Crofts, New York. [232]

Hinde, R. A. (1956). Ethological models and the concept of drive. *Br. J. Phil. Sci.* **6**, 321-331. [223]

Hinde, R. A. (1959). Unitary drives. *Anim. Behav.* **7**, 130-141. [207]

Hinde, R. A. (1960). Energy models of motivation. *Symp. Soc. exp. Biol.* **14**, 199-213. [215, 220, 221, 222, 223]

Hinde, R. A. (1969). (Ed.). "Bird Vocalisations". Their relations to current problems in biology and psychology. Cambridge University Press.

Hinde, R. A. (1970). Animal Behaviour. A synthesis of ethology and comparative psychology. 2nd Ed. McGraw-Hill, New York.
[25, 26, 125, 164, 187, 191, 200, 206, 212, 221, 223, 226]

Hinde, R. A. and Steel, F. A. (1966). Integration of the reproductive behavior of female canaries. *Symp. Soc. exp. Biol.* **20**, 401-426. [225]

Holst, E. von (1954). Relations between the central nervous system and the peripheral organs. *Br. J. Anim. Behav.* **2**, 89-94. [193, 196, 197, 205]

Holst, E. von, and Mittelstaedt, H. (1950). Das Reafferenzprinzip. *Naturwissenschaften* **37**, 464-476. [26, 193, 195, 196]

Holst, E. von and Saint Paul, V. von (1963). On the functional organisation of drives. *Anim. Behav.* **11**, 1-20. [224, 225, 236]

Hooker, T. and Hooker, B. I. (1969). Duetting. In Hinde (1969). [183]

Horn, G. (1967). Neuronal mechanisms of habituation. *Nature* **215**, 707-711. [200]

Howard, I. P., Craske, B. and Templeton, W. B. (1965). Visuo-motor adaptation to discordant ex-afferent stimulation. *J. exp. Psychol.* **70**, 189-191. [200]

Howard, I. P. and Templeton, W. B. (1966). "Human Spatial Orientation". John Wiley and Sons, New York and London. [17, 200]

Huggins, W. H. and Entwistle, D. R. (1968). "Introductory Systems and Design". Blaisdell Publishing Co., Waltham, Mass. [12, 61]

Hughes, M. T. G. (1969). Identification techniques. *In* "Modern Control Theory and Computing" (Bell, D. and Griffin, A. W. J., eds), pp. 138-163.

Hull, C. L. (1943). "Principles of Behaviour". Appleton-Century-Crofts, New York. [184, 185, 210, 211, 223]

Hull, C. L. (1952). "A Behaviour System". Yale University Press, New Haven. [185, [185, 235]

Hutt, S. J. and Hutt, C. (1970). "Direct Observation and Measurement of Behaviour". Charles Thomas. [124]

Iersel, J. J. van and Bol, A. C. A. (1958). Preening of two tern species. A study of displacement activities. *Behaviour* **13**, 1-88. [227, 236]

Jacobs, O. L. R. (1967). "An Introduction to Dynamic Programming". The theory of multistage decision processes. Chapman and Hall, London. [231]

James, W. (1890). "The Principles of Psychology". Macmillan, London. [17]

Kalman, R. E. (1960). On the general theory of control systems. *In* "Proceedings IFAC Moscow Congress", 1, 481-492. Butterworth, London. [141]

Kalman, R. E. (1963). Mathematical description of linear dynamical systems. *J. Soc. ind. appl. Matt.* (ser. A, Control) 1, 152-192. [145]

Kalman, R. E. (1968). New developments in systems theory relevant to biology. *In* "Systems Theory and Biology" (Mesarovic, M. D., ed.). Springer-Verlag, Berlin. [145, 146]

Kaufman, A. (1968). "The Science of Decision-making". World University Library. Weidenfeld and Nicolson, London. [231]

Kennedy, J. S. (1954). Is modern ethology objective? *Br. J. Anim. Behav.* 2, 12-19. [221]

Kissileff, H. R. (1969). Food-associated drinking in the rat. *J. comp. physiol. Psychol.* 65, 284-300. [127, 227, 235]

Korn, G. A. (1966). "Random-process Simulation and Measurements". McGraw-Hill, New York. [154, 159, 162]

Konishi, M. and Nottebohm, F. (1969). Experimental studies in the ontogeny of avian vocalisations. In Hinde (1969). [182, 183]

Koshikawa, S. and Suzuki, K. (1968). Study of osmoregulation as a feedback system. *Med. biol. Eng.* 6, 149-158. [69]

Kreindler, G. and Sarachik, P. E. (1964). On the concepts of controllability and observability of linear systems. *IEEE. Trans. Auton. Control.* AC-9, 129-136. [141]

Krendel, E. S. and McRuer, D. T. (1960). A servo-mechanisms approach to skill development. *J. Franklin Inst.* 269, 24-42. [110]

Lee, Y. W. (1960). "Statistical Theory of Communication". John Wiley and Sons, New York and London. [119]

Le Magnen, J. (1952). Les phenomenes olfacto-sexuels chez le rat blanc. *Archs Sci. physiol.* 6, 295-331. [187]

Le Magnen, J. (1968). Eating rate as related to deprivation and palatability in normal and hyperphagic rats. Reported at Third International Conference on the Regulation of Food and Water Intake. [68]

Le Magnen, J. and Tallon, S. (1966). La periodicite spontanée de la prise d'aliments ad libitum du rat blanc. *J. Physiol.* (Paris) 58, 323-349. [125, 127]

Lehrman, D. S. (1959). Hormonal responses to external stimulation in birds. *Ibis* 101, 478-496. [183, 225]

Lehrman, D. S. (1961). Gonadal hormones and parental behaviour in birds and infra-human mammals. *In* "Sex and Internal Secretion" (Young, W. C., ed.). Williams and Wilkins, Baltimore. [212, 224]

Leong, C. Y. (1969). The quantitative effect of releasers on the attack readiness of the fish *Haplochromic burtoni* (Cichlidae, Pisces). *Z. vergl. Physiol.* 65, 29-50. [190]

Lewin, K. (1936). "Principles of Topological Psychology". McGraw-Hill, New York. [231]

Lewin, K. (1943). Defining the 'field at a given time'. *Psychol. Rev.* 50, 288-290. [231]

Lewin, K., Dembo, Tamara, Festinger, L. and Sears, S. (1964). Level of aspiration. *In* "Personality and the Behaviour Disorders" (Hunt, J. M. V., ed.), Vol. 1. Ronald Press, New York.   [232]

Logan, F. A. (1960). "Incentive: How the Conditions of Reinforcement Affect the Performance of Rats". Yale University Press, New Haven.
                                                                [233, 234, 235]

Logan, F. A. (1964). The free behaviour situation. Nebraska Symposium on Motivation (Levine, D., ed.), Vol. XII, pp. 99-129.   [226, 233, 234, 235]

Logan, F. A. (1965a). Decision making by rats: delay versus amount of reward. *J. comp. physiol. Psychol.* **59**, 1-12.   [235]

Logan, F. A. (1965b). Decision making by rats: uncertain outcome choices. *J. comp. physiol. Pyschol.* **59**, 246-251.   [236]

Lorens, C. S. (1964). "Flowgraphs for the Modeling and Analysis of Linear Systems". McGraw-Hill, New York.   [122]

Lorenz, K. (1937). Uber die Bildung des Instinktbegriffes. *Naturwissenschaften* **25**, 289-300, 307-318, 324-331.   [220]

Lorenz, K. (1950). The comparative method in studying innate behaviour patterns. *Symp. Soc. exp. Biol.* **4**, 221-268.
                                   [181, 186, 206, 214, 215, 220, 221, 222, 223, 236]

Luce, R. D. (1959). "Individual Choice Behaviour". John Wiley and Sons, New York and London.   [118]

MacCorquodale, K. and Meehl, P. E. (1948). "On a Distinction Between Hypothetical Constructs and Intervening Variables". *Psychol. Rev.* **55**, 95-107.   [137]

Macfarlane, A. G. J. (1964). "Engineering Systems Analysis". G. G. Harrap, London.   [144]

Machin, K. E. (1964). Feedback theory and its application to biological systems. *Symp. Soc. exp. Biol.* **18**, 421-445.

Macmillan, R. H. (1962). "Non-linear Control Systems Analysis". Pergamon Press, Oxford.   [149]

Marler, P. and Hamilton, W. J. III. (1966). "Mechanisms of Animal Behaviour". John Wiley and Sons, New York and London.   [96, 97]

McDougall, W. (1923). "An Outline of Psychology". Methuen, London.
                                                                [215, 220, 222]

McFarland, D. J. (1964). Interaction of hunger and thirst in the Barbary dove. *J. comp. physiol. Psychol.* **58**, 174-179.   [69, 116, 160, 225]

McFarland, D. J. (1965a). The effect of hunger on thirst motivated behaviour in the Barbary dove. *Anim. Behav.* **13**, 286-292.   [69, 116, 240]

McFarland, D. J. (1965b). Hunger, thirst and displacement pecking in the Barbary dove. *Anim. Behav.* **13**, 292-300.   [228]

McFarland, D. J. (1965c). Control theory applied to the control of drinking in the Barbary dove. *Anim. Behav.* **13**, 478-492.   [55, 56, 62, 69, 169, 209]

McFarland, D. J. (1965d). Flow graph representation of motivational systems. *Br. J. math. stat. Psychol.* **18**, 1-9.   [61]

McFarland, D. J. (1966a). The role of attention in the disinhibition of displacement activity. *Q. J. exp. Psychol.* **18**, 19-30.   [225, 227]

McFarland, D. J. (1966b). On the causal and functional significance of displacement activities. *Z. Tierpsychol.* **23**, 217-235.   [201, 223, 225]

McFarland, D. J. (1966c). A servoanalysis of some effects of effort on response rate. *Br. J. math. stat. Psychol.* **19**, 1-13.   [138, 139, 140, 213, 214]

McFarland, D. J. (1967). Phase relationships between feeding and drinking in the Barbary dove. *J. comp. physiol. Psychol.* **63**, 208-213.   [69, 116, 117]

McFarland, D. J. (1969a). Mechanisms of behavioural disinhibition. *Anim. Behav.* **17**, 238-242.   [127, 160, 227, 233, 235]

McFarland, D. J. (1969b). Separation of satiating and rewarding consequences of drinking. *Physiol. Behav.* **4**, 987-989.   [68, 110]

McFarland, D. J. (1970a). Recent developments in the study of feeding and drinking in animals. *J. psychosom. Res.* **14**, 229-237.   [41, 102]

McFarland, D. J. (1970b). Adjunctive behaviour in feeding and drinking situations. *Rev. Comp. Anim.* **4**, 64-73.   [178, 227, 228, 233, 249, 251]

McFarland, D. J. (1970c). Behavioural aspects of homeostasis. "Advances in the Study of Behaviour" (Lehrman, D., *et al.*, eds), Vol. III, pp. 1-26. Academic Press, New York and London.
                              [144, 207, 209, 212, 213, 214, 215, 218, 220]

McFarland, D. J. (1972). A study of thirst as a motivational system in the Barbary dove *(Streptopelia risoria)*. *Anim. Behav. Monogr.* (In preparation).   [46, 239, 240, 241, 242, 243, 244, 245]

McFarland, D. J. and L'Angellier, A. B. (1966). Disinhibition of drinking during satiation of feeding behaviour in the Barbary dove. *Anim. Behav.* **14**, 463-467.   [227]

McFarland, D. J. and Baher, E. (1968). Factors affecting feather posture in the Barbary dove. *Anim. Behav.* **16**, 171-177.   [38, 39, 40, 50]

McFarland, D. J. and Budgell, P. (1970a). The thermoregulatory role of feather movements in the Barbary dove *(Streptopelia risoria)*. *Physiol. Behav.* **5**, 763-771.   [48, 49, 50, 51, 52]

McFarland, D. J. and Budgell, P. (1970b). Determination of a behavioural transfer function by frequency analysis. *Nature* **226**, 966-967.
                              [49, 75, 77, 78, 79, 153]

McFarland, D. J. and McFarland, F. J. (1968). Dynamic analysis of an avian drinking response. *Med. biol. Engng.* **6**, 659-668.
                  [41, 63, 65, 68, 69, 70, 111, 112, 168, 169, 177, 235]

McFarland, D. and McFarland, F. J. (1969). "An Introduction to the Study of Behaviour". Blackwell, Oxford.   [198]

McFarland, D. J. and McGonigle, B. (1967). Frustration tolerance and incidental learning as determinants of extinction. *Nature* **215**, 786-787.

McFarland, D. J. and Lloyd, I. H. (1971a). Determination of a behavioural impulse response by a stochastic identification technique. *Nature.* (In press).   [163]

McFarland, D. J. and Lloyd, I. H. (1971b). Time-shared feeding and drinking. (In preparation).

McFarland, D. J. and Perinchief, P. (1971). Electrolyte concentration in relation to hunger and thirst in the Barbary dove *(Streptopelia risoria)*. (In preparation).   [62]

McFarland, D. J. and Rolls, B. J. (1971). Suppression of feeding by intracranial injections of angiotensin. *Nature.* (In press).   [161]

McFarland, D. J. and Wright, P. J. (1969). Water conservation by inhibition of food intake. *Physiol. Behav.* **4**, 95-99.   [56, 69, 224]

McGonigle, B., McFarland, D. J. and Collier, P. (1967). Rapid extinction following drug inhibited incidental learning. *Nature* **214**, 531-532.

Machin, K. E. (1964). Feedback theory and its application to biological systems. *Symp. Soc. exp. Biol.* **18**, 421-445. [101]

Mason, S. J. (1953). Feedback theory—some properties of signal-flow graphs. *Proc. I.R.E.* **41**, 1144-1156. [12]

Mason, S. J. (1956). Feedback theory—further properties of signal-flow graphs. *Proc. I.R.E.* **44**, 920-926. [12]

Mayer, J. (1952). The glucostatic theory of regulation of food intake and the problem of obesity. *Bull. New Eng. med. Cen.* **14**, 43-49. [56]

Meehl, P. E. and McCorquodale, K. (1948). On a distinction between hypothetical constructs and intervening variables. *Psychol. Rev.* **55**, 95-107. [222]

Merton, P. A. (1961). The accuracy of directing the eyes and hand in the dark. *J. Physiol.* **156**, 555-577. [18]

Merton, P. A. (1964). Human position sense and sense of effort. *In* "Homeostasis and Feedback Mechanisms". *Symp. Soc. exp. Biol.* **18**, 387-400. [17]

Messenger, J. B. (1968). The visual attack of the cuttlefish *Sepia officinalis*. *Anim. Behav.* **16**, 342-369. [18]

Meyer, D. R. (1952). The stability of human gustatory sensitivity during changes in time of food deprivation. *J. comp. physiol. Psychol.* **45**, 373-376. [187]

Milhorn, H. T. Jr. (1966). "The Application of Control Theory to Physiological Systems". Saunders Press, Philadelphia.
[13, 44, 53, 54, 70, 71, 90, 100, 138, 169]

Miller, G. A. and Frick, F. C. (1949). Statistical behaviouristics and sequences of responses. *Psychol. Rev.* **56**, 311-328. [116]

Miller, N. E. and Kessen, M. L. (1952). Reward effects of food via stomach compared with those of food via mouth. *J. comp. physiol. Psychol.* **45**, 555-564. [68]

Miller, N. E., Sampliner, R. I. and Woodrow, P. (1957). Thirst reducing effects of water by stomach fistula vs water by mouth measured by both a consummatory and instrumental response. *J. comp. physiol. Psychol.* **50**, 1-5. [177]

Milsum, J. H. (1966). "Biological Control Systems Analysis". McGraw-Hill, New York. [13, 22, 54, 61, 71, 72, 73, 81, 82, 83, 85, 86, 90, 94, 102, 105, 107, 119, 124, 144, 152, 153, 237]

Milsum, J. H. (1968). Optimization aspects in biological control theory. *In* "Advances in Biomedical Engineering and Medical Physics" (Levine, S. N., ed.) **1**, 243-278.

Mittelstaedt, H. (1957). Prey capture in mantids. *In* "Recent Advances in Invertebrate Physiology—A Symposium" (Scheer, B. T., ed.). Eugene, University of Oregon Publications. [19, 27, 28, 29, 32, 33]

Moroney, M. J. (1951). "Facts from Figures". Pelican. [118, 124]

Mulligan, J. A. and Olsen, C. (1969). Communication in canary courtship calls. In Hinde (1969). [183]

Nachman, M. and Pfaffman, C. (1963). Gustatory nerve discharge in normal and sodium-deficient rats. *J. comp. physiol. Psychol.* **56**, 1007-1011. [187]

Nakayama, T., Hammel, H. T., Hardy, J. D. and Eisenman, J. S. (1963). Thermal stimulation of electrical activity of single units of the preoptic region. *Am. J. Physiol.* **204**, 1122-1126.   [23]

Naslin, P. (1962). Computation of transients. *In* "Non-linear Control Systems Analysis" (Macmillan, ed.). Pergamon Press, Oxford.   [152]

Naslin, P. (1965). "The Dynamics of Linear and Non-linear Systems". Blackie, London.   [61, 152]

Notterman, J. M. and Mintz, D. E. (1965). "Dynamics of Response". John Wiley and Sons, New York and London.   [118]

Novin, D. (1962). The relation between electrical conductivity of brain tissue and thirst in rat. *J. comp. physiol. Psychol.* **55**, 145-154.   [177]

Oatley, K. (1967). A control model for the physiological basis of thirst. *Med. biol. Engng.* **5**, 225-237.   [62, 69, 116, 160, 169, 175, 235]

Oatley, K. and Tonge, D. A. (1969). The effect of hunger on water intake in rats. *Q. J. exp. Psychol.* **21**, 162-171.   [116, 225]

Olson, H. F. (1958). "Solutions of Engineering Problems by Dynamical Analogies". Van Nostrand, Princeton.   [144]

Pace, W. H. (1961). An analogue computer model for the study of water and electrolyte flows in the extracellular and intracellular fluids. *I.R.E. Trans. Bio-Med. Electron.* BME-8(1).   [69, 169]

Pfaffman, C. and Bare, J. K. (1950). Gustatory nerve discharges in normal and adrenalectomised rats. *J. comp. physiol. Psychol.* **43**, 320-324.   [187]

Prime, H. A. (1969). "Modern Concepts in Control Theory". McGraw-Hill, New York.   [230, 231]

Rabbitt, P. M. A. (1968). Three kinds of error-signalling responses in a serial choice task. *Q. J. Exp. Psychol.* **20**, 179-188.   [118]

Raynor, J. O. (1969). Future orientation and motivation of immediate activity: An elaboration of the theory of achievement motivation. *Psychol. Rev.* **76**, 606-610.   [232]

Reeve, E. B. and Kulhanek, L. (1967). Regulations of body water content: A preliminary analysis. *In* "Physical Basis of Circulatory Transport: Regulation and Exchange (Reeve and Guyton, eds). Saunders Press, London.   [69, 235]

Rock, I. and Harris, C. S. (1967). Vision and Touch. *Scient. Am.* **216**, 96-104.
                                                                              [198]

Rosen, R. (1967). "Optimality Principles in Biology". Butterworth, London.
                                                                         [237, 252, 256]

Rowell, C. H. F. (1961). Displacement grooming in the chaffinch. *Anim. Behav.* **9**, 38-63.   [125, 227, 228]

Schultz, D. G. and Melsa, J. L. (1967). "State Functions and Linear Control Systems". McGraw-Hill, New York and London.
                          [45, 141, 142, 145, 146, 149, 151, 152, 217, 218, 219, 231]

Seitz, A. (1940). Die Paarbildung bei einigen Cichliden. *Z. Tierpsychol.* **4**, 40-84.   [190]

Sensicle, A. (1968). *In* "Introduction to Control Theory for Engineers". Blackie, London. [100, 103, 107, 148]

Sevenster, P. A. (1961). A causal analysis of a displacement activity (fanning in *Gasteroteus aculeatus* L.) *Behaviour Suppl.* **9**, [207, 227, 228]

Sevenster-Bol, A. C. A. (1962). On the causation of drive reduction after a consummatory act. *Archs néerl. Zool.* **15**, 175-236. [206]

Sherrington, C. S. (1906). "The Integrative Action of the Nervous System". Yale University Press. [37, 225]

Sherrington, C. S. (1918). Observations on the sensual role of the proprioceptive nerve-supply of the extrinsic ocular muscles. *Brain* **41**, 332-343. [17, 197]

Siegel, S. (1956). "Nonparametric Statistics for the Behavioural Sciences". McGraw-Hill, New York. [36, 37]

Skinner, B. F. (1938). "The Behaviour of Organisms: An Experimental Analysis". Appleton-Century-Crofts, New York. [137]

Skinner, B. F. (1950). Are theories of learning necessary? *Psychol. Rev.* **57**, 193-216. [137]

Sokolov, E. N. (1960). Neuronal models and the orienting reflex. *In* "The Central Nervous System and Behaviour" (Brazier, M. A. B., ed.). Macy Foundation, New York. [200]

Spence, K. W. (1956). "Behavior Theory and Conditioning". Yale University Press, New Haven. [235]

Spiegel, M. R. (1961). "Theory and Problems of Statistics". Schaum Publishing Co., New York. [118, 124]

Stark, L. (1959). Stability, oscillations, and noise in the human pupil servomechanism. *Proc. I.R.E.* **47**, 1925-1936. [87, 92, 138]

Stark, L. (1962). Biological rhythms, noise, and asymmetry in the pupil-retinal control system. *Ann. N.Y. Acad. Sci.* **98**, 1096-1108. [94]

Stark, L. (1968). "Neurological control systems. Studies in Bioengineering". Plenum Press, New York.
[51,54, 88, 91, 92, 93, 94, 130, 1, 132, 133, 154, 156, 157, 158]

Stark, L. and Sherman, P. M. (1957). A servoanalytic study of the consensual pupil reflex to light. *J. Neurophysiol.* **20**, 17-26. [91]

Stark, L. and Young, L. R. (1964). Defining biological feedback control systems. *A. N.Y. Acad. Sci.* **117**, 426-442. [137]

Stark, L., Vossius, G. and Young, L. R. (1962). Predictive control of eye tracking movements. *I.R.E. Trans. on Human Factors in Electronics*, HFE-3(2), 52-57. [138]

Stellar, E. and Hill, J. H. (1952). The rat's rate of drinking as a function of water deprivation. *J. comp. physiol. Psychol.* **45**, 96-102. [175]

Stern, H. J. (1944). Simple method for early diagnosis of abnormalities of pupillary reaction. *Br. J. Opthalmol.* **28**, 276-278. [93]

Stevens, S. S. (1951). Mathematics, measurement and psychophysics. *In* "Handbook of Experimental Psychology" (Stevens, S. S., ed.). John Wiley and Sons, New York and London. [35, 37]

Stewart, C. A. and Atkinson, R. (1967). "Basic Analogue Computer Techniques". McGraw-Hill, New York. [168]

Thomas, D. W. and Mayer, J. (1968). Meal taking and regulation of food intake by normal and hypothalamic hyperphagic rats. *J. comp. physiol. Psychol.* **66**, 642-653. [125, 235]

Thurstone, L. L. (1919). The learning curve equation. *Psychol. Monogr.* **26**, 1-37.   [116]

Tinbergen, N. (1951). "The Study of Instinct". Clarendon Press, Oxford.
                                              [205, 215, 220, 221, 222, 223]

Toates, F. M. and Oatley, K. (1970). Computer simulation of thirst and water balance. *Med. biol. Engng.* **8**, 71-87.
                     [62, 69, 70, 170, 172, 173, 174, 175, 176, 177, 178, 235, 239]

Tugendhat Gardner, B. (1964). Hunger and sequential responses in the hunting behaviour of Salticid spiders. *J. comp. physiol. Psychol.* **58**, 167-173. [236]

Tweel, L. H., van der (1969). Designated discussion, on J. J. Denier van der Gon and G. H. Wieneke, The concept of feedback in motorics against that of preprogramming. *In* "Biocybernetics of the Central Nervous System" (Proctor, L. D., ed.). Little, Brown & Co., Boston, Mass.   [115, 116]

Vowles, D. M. (1970). Neuroethology, Evolution and Grammar. *In* "Development and Evolution of Behaviour" (Aronson, I. R., Tobach, E., Lehrman, D. S. and Rosenblatt, J. S., eds). Essays in memory of T. C. Schneirla. W. H. Freeman & Co., Folkestone.   [124, 125]

Weiskrantz, L. (1968). Experiments on the r.n.s. (real nervous system) and monkey memory. *Proc. R. Soc. B.* **171**, 335-352.   [137]

Wells, D. A. (1967). "Theory and Problems of Lagrangian Dynamics". McGraw-Hill, New York.   [216, 256]

Wells, M. J. (1962). "Brain and Behaviour in Cephalopods". Heinemann, London.   [19]

Wiepkema, R. (1961). An ethological analysis of the reproductive behaviour of the bitterling. *Archs néerl. Zool.* **14**, 103-199.   [124]

Wiepkema, P. (1968). Behaviour changes in CBA mice as a result of one goldthioglucose injection. *Behaviour* **32**, 179-210.   [67]

Wilkins, B. R. (1966). Regulation and control in engineering. *In* "Regulation and Control in Living Systems" (Kalmus, H., ed.). John Wiley and Sons, New York and London.   [104]

Williams, B. J. and Clarke, D. W. (1968). *Control* **12**, 233-238.   [162]

Wolf, A. V. (1958). Osmometric analysis of thirst in man and dog. *Am. J. Physiol.* **161**, 75-86.   [175, 235]

Wright, P. and McFarland, D. J. (1969). A functional analysis of hypothalamic polydipsia in the Barbary dove *(Streptopelia risoria)*. *Physiol. Behav.* **4**, 877-883.   [56, 69]

Yates, A. J. (1962). "Frustration and Conflict". Methuen, London.   [207]

Yates, F. E., Brennan, R. D., Urquhart, J., Dallman, M. F., Li, C. C., Halpern, W. (1968). A continuous system model of adrenocortical function. *In* "Systems Theory and Biology" (Mesarovic, M. D., ed.). Springer-Verlag, Berlin.   [169, 212]

Young, L. R., Green, D. M., Elkind, J. L. and Kelly, J. A. (1963). Adaptive characteristics of manual tracking. Fourth National Symposium on Human Factors in Electronics, Washington, D.C.   [110, 138]

Young, W. C. (1961). The hormones and mating behaviour. *In* "Sex and Internal Secretion" (Young, W. C., ed.). Williams and Wilkins, Baltimore.   [212]
Zeigler, H. P. (1964). Displacement activity and motivation theory. A case study in the history of ethology. *Psychol. Bull.* **61**, 362-376.   [223]

Subject Index

Feedback—*cont.*
  discrepancy, 201
  fictitious, 20
  mechanism, 1, 20, 21, 138-140, 206
  oral, 63, 66-69, 110, 177, 179, 214, 248
  positive, 21, 66-70, 203, 214, 248
  proprioceptive, 17, 18, 98, 183, 197
  resistance, 167
  visual, 17-19, 27-34, 98
Feed-forward, 102
Feeding, 19, 56, 102, 116-118, 125-129, 160-162, 209, 224-226, 237-256
Final common path, 224-226
Fistulae, 68, 110
Fixation deficit, 30-33, 98
Flow, generalized, 4, 144, 210-225, 242
Flow graph, 10-12, 27, 30, 45, 58-61, 166, 168
Fly, 27, 195-196
Flying, 154-156
Fourier transform, 130, 135
Fowl, 62
Freedom, degrees of, 217
Frequency analysis, 43, 73, 78, 90-95, 133, 137, 153, 154, 159
  break, 80
  histogram, 119, 120, 125
  natural, 80-84, 106, 107, 111
Frustration, 207, 213, 251
Function of frequency, 43, 71-74
  of time, 7, 35, 37, 41, 43, 180
Functional optimization, 249-256

Gain, 92-94, 107, 111, 112, 166, 167
Gaussian distribution, 124, 131
Generalized capacitance, 211-225, 253-255
  coordinates, 216-219, 254
  displacement, 144, 211-225
  drive, 185-188, 213, 223
  effort, 4, 143, 210-225, 245
  flow, 4, 143, 210-225, 242
  inductance, 211-225
  momentum, 144, 211-225
  parameters, 4, 209-225
  variables, 4, 144, 209-225

Genetics, 35
Grammar, 124
Guppy, 124, 188
Gut, 58-70

Habit strength, 185-188
Hamilton's Principle, 256
Hardware dependence and independence, vi, 221
Helmholtz concentric ring, 136
Herring gull, 191, 204, 251, 252
Heterogenous summation, 190
Holonomic constraints, 217
Homeostasis, 1, 56, 102, 138, 169-179, 208, 209, 212, 237
Hormone, 55, 56, 69, 175, 183, 212, 213, 224
Hullian system, 184-188, 207, 211, 223, 235
Human operator, 104-110
Huygens, 1
Hypothalamus, 23, 51, 55-57, 102
Hysteresis, 177, 239

Impulse input, 47, 53
  response, 47, 83, 149, 153, 156-162
Incentive, 185, 213, 233, 236
Incompatibility, 37-40
Indentification, 152-165
Indifference functions, 236
Inductance, 3, 144, 211-225
Inertia, 10, 41, 222
Inflow theory, 197
Initial conditions, 9, 44, 58-61, 141, 153
Injury, 180, 183
Input, periodic, 43, 71, 75, 83
  variables, 7, 9, 26, 43
Integral control, 107, 108
Integrations, 44, 210, 217-219
Integrators, 44, 45, 56-70, 143, 183, 206
Intensity, 119, 125-127
Isoclines, 147-149
Isomorphism, 35, 37

Joint receptors, 18, 21